DISCARDED

HD
6509
M6
A3

# Struggle in the Coal Fields

## The Autobiography of Fred Mooney

Secretary-Treasurer, District 17
United Mine Workers of America

**Edited by J. W. Hess**

LIBRARY OF
LAMAR STATE COLLEGE OF TECHNOLOGY

↑ 935745

WEST VIRGINIA UNIVERSITY LIBRARY
MORGANTOWN
1967

Copyright © 1967
WEST VIRGINIA UNIVERSITY FOUNDATION
Morgantown, West Virginia

*Library of Congress Catalog Card Number 67-30927*

McClain Printing Company
Parsons, West Virginia

## CONTENTS

| | |
|---|---|
| Introduction | vii |
| Autobiography of Fred Mooney | 1 |
| Illustrations, between pages | 88-89 |
| Map, between pages | 96-97 |
| Footnotes | 166 |
| Bibliography | 174 |
| Index | 181 |

The publication of this book was
made possible by a grant from the

**WEST VIRGINIA UNIVERSITY
FOUNDATION**

# INTRODUCTION

Fred Mooney, secretary-treasurer of District 17, United Mine Workers of America from 1917 to 1924, had worked in the coal mines of Kanawha County since 1902. District 17 included all of West Virginia north and west of Charleston with the exception of eastern Preston County and the northern panhandle.

As an active member of the union and then as an elected officer, Mooney was a participant in much of the labor unrest in West Virginia during the first three decades of this century. Framed as an autobiography, his story covers little of his private life. He tells of his self-education, his experience as a working man, and his activity for the union. His narrative is published as a document for the social record, rather than as a disclosure of sensational events not previously reported. The reader will readily detect, but is nonetheless advised in advance, that Mooney was obviously biased in favor of the mine worker and against the coal operators. He accepted the versions of events that circulated among the miners, some of which were surely exaggerated. After checking newspapers and other versions of the events, however, I have concluded that from the union viewpoint it is not an irresponsible account, although some interpretations might be questioned and Mooney does not pretend to be impartial.

The United Mine Workers of America made efforts to organize in West Virginia shortly after the union was formed in 1890. At that time West Virginia's coal production was small compared to the national total, which was predominantly mined in the Central Competitive Field (western Pennsylvania, Ohio, Indiana, and Illinois). West Virginia's relationship to the other coal fields, however, was an important aspect of the story. In 1897, the operators of the Central Competitive Field recognized the union and, at about the same time, became intensely aware of increasing competition from West Virginia. West Virginia operators could sell in both eastern and mid-western markets and found work stoppages in other areas particularly to their advantage.

The operators in the central fields undoubtedly pressured the union to organize West Virginia. Operators in the latter state openly declared their intention to operate in defiance of the union, pleading that their workers were better paid than union labor, that their shipping and operating costs were greater, and that they could not afford the costs which would result from recognition of the union. The union could not hope to maintain its position where it had been recognized unless it could organize West Virginia, where production increased even more rapidly after the opening of the Logan field in the southern part of the state in 1904. With regard to the situation in West Virginia, one must also consider a tradition of weak law enforcement and increasing disillusionment on the part of many who had been attracted to the mines from the farms of Appalachia, the deep South, and the nations of Europe. Violent eruptions might then be expected.

The U. M. W.'s earliest efforts in West Virginia were completely frustrated by court injunctions and by special police and detectives hired by the operators. During the national strike of 1902, the union made progress in the Kanawha fields and in a few mines in other areas. The union lost ground, however, between 1903 and the Paint and Cabin Creek outbreak of 1912-13. Strikes and lockouts during the latter period led to violence, numerous arrests, and extended periods of martial law. When the disorder was ended by compromise, the union was precariously established in the Kanawha Valley. The organization continued to advance in West Virginia during World War I, when production expanded rapidly and the federal government, applying wartime controls, insisted on collective bargaining.

President of District 17 during Mooney's tenure as secretary-treasurer was C. Frank Keeney. Their first two years in office were favored by wartime prosperity and an atmosphere of national support. Membership was increased dramatically and by 1919 organization of the fields north of Charleston and the

Kanawha area was virtually complete. Even during this favorable period, the union was excluded from the newer but very productive coal fields of the southern part of the state.

A reactionary mood swept the nation following World War I. The "red scare" of 1919-20, the upsurge of the Ku Klux Klan, and a strong anti-union trend were features of the period. In the drive against the union, operators asked for and often received injunctions prohibiting organizational activity on or around their property. These injunctions had been upheld by the United States Supreme Court in a case involving the Hitchman Coal and Coke Company's operations in northern West Virginia. The case originated in 1907 and was decided by the Court in 1917; it had had little effect in the latter year because of wartime regulations but was influential during the post-war period.

The anti-union movement, which was often promoted as "the American plan," was more immediately successful in West Virginia where organization had never been complete and self-sustaining. Drastic post-war fluctuations in the market for coal also contributed to conditions in the West Virginia fields which led to the miners' marches of 1919 and 1921, the Matewan Massacre of 1919, and other serious disorders. Martial law was declared in August, 1920, and federal troops occupied portions of the state between August 28 and November 4, and between November 28 and February 15, 1921. Federal troops were again dispatched on September 4, 1921. Preservation of order often involved the protection of non-union workers from interference by strikers, and operators were actually able to increase production during a strike in the critical southern part of the state. While union leaders were imprisoned and tried for treason against the state following the armed march of 1921, membership declined rapidly. By 1924, membership in District 17 was about half what it had been in 1921; by 1928, the union was virtually non-existent in West Virginia. Mooney tells this story and adds information on the conflicts within the union which hampered its efforts. Keeney and Mooney resigned (or were removed) from leadership of the district in 1924. They thereafter unsuccessfully opposed John L. Lewis as he consolidated his control of the international union.

Keeney and Mooney were at times referred to, or even referred to themselves, as socialists. Class struggle, for them, was not a revolutionary theory but a fact of life. They were not members of a national or international conspiracy or socialist organization. Mooney confessed that he could not understand Karl Marx, and the Russian revolution had not occurred or had not been understood when their attitudes were formed and when they engaged in most of the labor activity covered in this manuscript.

In 1925, Mooney went to New Mexico, where he worked as a carpenter. In 1926, he was in Florida and was associated with a labor journal called the *St. Petersburg Union,* official organ of the St. Petersburg Central Labor Union and the Clearwater Building Trades Council. Most of this narrative through Chapter V was published in that weekly newspaper in somewhat different form.

Mooney moved from Florida to the Far West late in 1926. Between that date and 1934, when the autobiography ends, he did construction work in Texas, Arizona, and New Mexico; returned to Charleston to operate a restaurant between late 1927 and April, 1928; worked in Missouri, New Mexico, and Arizona in 1928-29; returned to Charleston in 1930; and worked in Texas and the Far West between April, 1931, and February, 1934, when he returned again to West Virginia. After 1934, Mooney worked for coal companies in the vicinity of Fairmont, Grafton, Morgantown, and Elkins, sometimes as superintendent and sometimes as section foreman and fire boss. He died of a self-inflicted gunshot wound on February 24, 1952, following a period of despondency. According to Fairmont newspapers, Mooney's suicide followed a "booby-trap" explosion in his home for which police assumed he was responsible.[1]

Most of the manuscript was written in the late 1920s or in the 1930s. Mooney revised it at least three times, making insertions but no significant changes. On one revision, when commercial publication was considered, names of persons and places were fictionalized; this version was not completed. During these revisions, others obviously assisted with such details as grammar and punctuation. I have eliminated some repetitive paragraphs, corrected punctuation, and restructured sentences when awkwardness created doubt or made reading difficult. Some detailed reports of isolated violence during the strikes have been omitted. A few portions of Chapter VII, on Mooney's Mexican trip in 1920, have been judged meaningless and not very interesting tourist reports and therefore were omitted. His journeys in search of employment in Chapters XI and XIII have been shortened somewhat for the same reason. Even with changes of this kind, the manuscript is essentially in Mooney's style, containing those things which he wanted to say with the tone he wanted to give them. My only insertions are in the form of footnotes, organized by chapter at the end of the text, when further information or supporting or contradictory references were considered worthwhile.

The original manuscript is now the property of the West Virginia University Foundation, Inc., and is available in the West Virginia University Library. This publication would not have been undertaken or completed without the encouragement and assistance of Robert F. Munn, director of libraries. Charles Shetler, at the time curator of the West Virginia Collection, originally acquired a copy of the autobiography from Mrs. Virginia R. Mooney. Edward M. Steel, associate professor of history, and Stokely B. Gribble, assistant director of libraries, offered valuable comments and suggestions. Mrs. Pauline T. Kissler, assistant in the manuscript section, contributed her knowledge of numerous sources. Mrs. Belva Atwood and Mrs. Martha Neville of the Collection staff typed the manuscript, and Joan Ellis, assistant curator, helped with the final proofreading.

J. W. H.

## Chapter I

I was born on January 23, 1888, in a log cabin on Davis Creek in Kanawha County, West Virginia. The house was a typical pioneer affair with two rooms on the ground floor. The top story extended across a porch in between the two rooms and the top part of the house was partitioned into three rooms; however, these rooms had very low ceilings and were only used for sleeping. A small lean-to or shed porch had been added to the south end of the house and this was generally referred to as the front.

My parents were of pioneer stock. My father worked in the mines, on small timber-cutting jobs, or any other employment at which a dollar or its equivalent in groceries could be earned. Wages were very low in those days and my father was lucky if he was paid as much as one dollar per day for 10 hours work. Had we not raised cows, pigs, and chickens and farmed several acres of ground in corn, sorghum-cane, and garden truck, we would have been confronted by the wolf at the door many times.

My mother was the daughter of a coal miner and in this environment no choice was presented to me other than timber or the coal mines. I attended public school until I was 18 years of age but went to work as a trapper boy in the mines when only 13. I worked during the off-school months and frequently over into the school term for one or two months in order to be able to purchase books and clothing for the school period.

As a trapper boy, I received 25 cents per day for a 10-hour shift. After being on the job for a short time, I was promoted to attend two trap doors and tend the fire in the ventilating furnace. For this work I was paid 75 cents per day. I imagined myself well on the road to wealth.

One of my earliest developments was an inquiring turn of mind. I wanted to know the whys and wherefores of everything. My father and mother were strict adherents to the church, being affiliated with what was then classified as the Missionary, or "Hard-shell," Baptists. Occasionally my mother would back-slide and when she re-affiliated with the church she would join a different denomination such as the Freewill, or "Soft-shell," Baptists. However, eventually she would drift back to the "Hard-shells" again. When it was at all possible for them to do so, my parents compelled me to attend church, Sunday school, and prayer meeting. At times this routine would become monotonous to me and I would rebel.

Preachers were inclined to flock to our home for week ends and on many occasions my two brothers and I stood in the background and watched the last piece of chicken and the last dumpling disappear down the neck of a preacher. My wishes concerning the future abiding place of the clergy would have shocked these pious gentlemen, had they been expressed aloud.

One Friday afternoon, near the finish line of one of the revivals that were held two or three times each year, my father said to me, "Son, we will have to wheel rock and sod tomorrow and dam up the creek out front so brother Pelly can be baptized on Sunday."

"Brother Pelly? You mean that man that cries all the time?" I inquired.

"Yes, but you must not speak of him in that way."

"Why not, Dad? He does cry, don't he?"

"Yes, he does," he replied with a show of impatience, "but the reason he cries is because he really has religion."

"Does really having religion cause people to cry when they preach, pray, sing, and when asking a blessing at the table?" I wanted to know.

"No, not all of them do that. Religion affects different people in different ways."

"Why can't he be baptized down in the swimmin' hole where the other people were baptized?"

"Too far away from the house and they are afraid he might take cold from being in the water."

"Why preacher Smith said Sunday week ago, when he baptized those people down at the swimmin' hole, that no one ever took cold from being in the water to be baptized in the name of the Lord."

Withering me with a rather stern look, he said, "You had better go do your chores now for we have a hard day before us tomorrow."

While I did the chores that evening, the thought kept running through my mind, "Why can't that preacher be baptized down in the swimmin' hole like the others were? Here we have to work all day gathering rock and sod to dam up a special hole for him. Just watch, they will all be here for dinner on Sunday, too." I retired that night still thinking of that dam with resentment.

Saturday the back-breaking job of building that dam was completed and when finished we had raised the water about one foot. We attended church that night and every bone and muscle in my body ached. During the services my father was compelled to prod me several times to keep me awake. I was always instructed to sit near the preacher, in the "Amen corner," and was told to "sit still and don't fidgit around." Services began with an opening hymn. Then followed two or three other hymns, a prayer, the reading of a chapter, another prayer; then a text was selected and the sermon began. The sermons were long and contained plenty of hell-fire and brimstone. Near the end of the services, brother Pelly, who was to be baptized next day, was called upon to explain why he was transferring from the Winebrennarian church[1] to the Missionary Baptist denomination.

"Well, many of you know Jeb Haskins and his wife. It was because of a remark that I made to brother Mounds relative to visiting sister Haskins while she was ill that caused the trouble."

"What was it that you said to the brother that was held against you?" inquired the minister.

"Well, as you know, sister Haskins has been seriously ill for some time and on a Sunday morning I started down to visit her. On the way I met brother Mounds as he returned from a visit to the Haskins!"

"Where have you been, brother Mounds?"

"I was calling on and sympathizing with sister Haskins."

"Did you take her anything to eat?"

"No, I didn't."

" 'Then,' I said, 'brother Mounds, there are better men in hell than you are.' That was my remark to the brother and they charged me with using profanity and taking the name of the Lord in vain. I had visited the ailing sister on several occasions and had taken her a small chicken, butter, fruits, and a loaf of bakers' bread each time. That is the reason I am here tonight and wish to be baptized tomorrow. If only praying with sick people and sympathizing with them when I know they are hungry and need food is religion, then I haven't and don't want any of it."

About one month after brother Pelly was baptized, a revival meeting was conducted and several new converts were admitted to the church. Arrangements were made for their baptism at a later date.

My principal responsibility consisted of doing chores and caring for the smaller children while my parents attended the meetings which were held at night. Occasionally I was permitted to attend.

Regular attendants at these meetings were three elderly women who were professional shouters. When a group of people, both old and young, reached that stage generally alluded to as "being under conviction," they were arranged along a mourners' bench in front of the pulpit. When excitement began to run high the pastor would nudge sister Mary, Alice, or Bertie and tell them to shout. Two of these women while going through the performance generally looked where they were going. But not sister Mary. When she shouted, she began by jumping up and down and her vertical momentum was always backwards. When she began to show signs of going into action my father would give me the sign to get behind Aunt Mary and prevent her from falling over one of the benches.

I do not wish to be misunderstood. These were a great people. It is true that they were intolerant in their attitude toward one another's religion, political views, and affiliations. However, they were charitable in the broad sense of the word. If someone were ill and the necessities of life began to run low, baskets of provisions began to move toward the unfortunate neighbor. The larder was kept replenished until the sick or afflicted was again able to provide for his own household. I pass lightly over these incidents in order to show the environment with which I was surrounded during my early years.

Often during my impressionable teens I almost convinced myself that I was under conviction. I did not go to the mourners' bench, yet I was always in close proximity. I did not reveal my state of mind to anyone, but after retiring for the night I prayed long drawn-out prayers. My parents often threatened me with the devil, telling me that if I did not cease being a community "black sheep" Satan would get me some day. One night in a dream Satan called for me. He was equipped with the proverbial horns, had shining eyes, and wore an iron overcoat. His overcoat opened in front in conventional style, but under the arms and running down the sides it was hinged. He walked in, picked me up, placed me inside his coat and closed the doors on me.

"Please do not take me," I pleaded. "There are boys down the road who are worse than I am. Get one of them this time, and I will do better." He unhinged his coat, set me down, and departed without comment concerning my promise. When he released me I awakened and was standing in the middle of the floor several feet away from my bed and had been in a night sweat. I was miserable and afraid.

"Did you see anyone in the house?" I inquired of my mother.

"No, son, you must have been dreaming."

Our home was located on the northern edge of a 7,000-acre virgin forest. The hand of man had never wielded the axe and crosscut here. This tract of timber was purchased by a stave and lumber company shortly after the close of the Spanish-American War. The entire community was elated at the coming of the mills. "There will be plenty of work," was the general expression. The rejoicing at the coming of the mills soon turned to chagrin when it was learned what the wages were going to be. Common labor was $1.00 per day for 10 hours; $1.50 was the highest rate except for woods bosses, who were paid $2.00 per day. The latter were looked up to as the elite of the camp. Boys who were hired to stack staves, wheel sawdust, and carry out refuse were paid $.25, $.50, and, in rare cases, $1.00 for a day's work.

I was 12 years old in January of the year the operation began in our area. When school was out I said to my father, "Dad, I want to go to work."

"You'll do nothing of the kind," he said, "you'll take care of the garden and the field this summer."

So I moped and pouted for several days. Fences began to go down to make room for haulage ways; apple trees were used for hitching posts, and the horses and mules gnawed the bark from them. It developed that we were going to be confined to a very small garden insofar as tilling the soil was concerned. Eventually my father consented for me to work at stacking staves.

The rejoicing at the coming of the mills soon turned into bitterness and chagrin at the devastation wrought by the cutting and marketing of the timber.

I stacked staves, wheeled out sawdust, and did odd jobs through 1901 and into 1902. It was a fight every inch of the way, for anyone who knows anything about a lumber camp or stave camp will understand that to stick around one has to "hold his own." I was not yet physically able to take care of myself among the crowd, but several of them soon learned to their sorrow that I could toss a pebble with devastating effect and most of them let me alone. However, on the job where I was wheeling sawdust there were two men who seemed to derive an immense satisfaction from teasing boys. They both started in on me. One day I threw a rock at one of them, missed him, and hit the governors on the steam engine. The rock broke the governors and caused the mill to be closed down for several hours. For this act I was discharged. The man who was responsible for my unforgivable deed was not even reprimanded.

When leaving work he was compelled to pass by my home and walk across a foot-log about eight feet above the creek. Securing a good sized pebble, I waited until he reached the center of that foot-log and I knocked him into the creek. He came to the house wet from head to foot and called to my father, "Watt, you will have to do something with that boy of yours or I am going to have him arrested."

I had already related to my father what had happened that afternoon, and he walked out to the man and said, "And just what am I going to do with you? You have meddled with that boy now until you have caused him to be discharged. Now, get out of here before someone carries you out." These flare-ups from my father always amused me for he was so small. Even in the prime of his life he never weighed more than 132 pounds.

From him I inherited my tendency to bet on the tossing of a pebble when in a pinch. He could throw a stone with uncanny speed and accuracy. Someone circulated a story concerning his pebble tossing. They said they met him coming out of the woods with three gray squirrels in his hand; yet, he carried no gun. They inquired, "Watt, where is your gun?"

"Didn't have any gun; killed them with rocks," he replied. "See that big one there (holding up the squirrel to view)? Had to throw at him twice." He was easy going, quiet and under ordinary circumstances had little to say. Relative to the rock I threw at the man, he said, "Better be careful who you throw rocks at. It will get you into trouble."

Sunday school appealed to me, for here knowledge could be obtained. I wanted to know the Bible. This desire became an obsession with me and I was not satisfied until I had read it several times. When I asked a question in Sunday school or out concerning a passage of the Bible that I did not understand, sometimes I got an answer and occasionally an excuse. "There are portions of that scripture which are not supposed to be understood," I was told many times. This puzzled me. Here was a book which laid down a routine of life to which we were supposed to conform, yet I was told that some of it was not understandable. I did not ask why this time.

I went to Bible history, *Bible Scenes and Studies,* and to the works of Dr. Talmadge, Dwight L. Moody, Sam Jones and others to learn all I could concerning the Bible. To a seeker after knowledge one certain line of thought eventually becomes monotonous. So it was with me.

I next read *Heralds of the Morning,* the author of which I do not remember. This book interpreted many of the scriptures in a more enlightening manner. Upon the recommendation of a friend I read *The Jungle* and *Prince Hagen* by Upton Sinclair, *The Iron Heel* by Jack London, and *The Struggle for Existence* by Walter Thomas Mills.[2] At this stage I imagined myself ready to make speeches but soon realized that I was crude, rough, and raw. Something must be done to correlate my ideas and govern my expression.

I secured a copy of *The Art of Lecturing* by Arthur M. Lewis, read it and went to the woods. I would locate a quiet spot, scout around to see if there was anyone in the immediate vicinity, then, if the coast was all clear, I would mount a stump, log, or rock, clear my throat, and address the audience. I would imagine myself sitting in the audience as a listener. For a long time after beginning this practice I would question myself at the conclusion of the spiel. Later I would heckle myself during the speech, call myself a liar, tell myself "you don't know what you are talking about." When I left one of these lonely spots I had picked about all the flaws in my speech that any ordinary person would be able to reason out, or at least I thought so. In my conversations with people, I was too insistent and arrogant and few converts flocked to my banner. I did not seek advice from anyone. I wanted to reason things out for myself. In later years this attitude led me into strange fields and circumstances from which extrication was often difficult, painful, and fraught with much worry. However, the pursuance of this policy bore fruit, for it developed a tendency to rely upon myself. My maternal grandfather often said to me, "Listen, son, when you want to do something, go and do it yourself; take no friends or partners along, for nothing is a secret if it is known by more than one person."

In 1902, the year I went to work at the mine, the miners' union was established in the community. In order to be regular I joined.[3] I did not know what for nor why but everyone was "jinin" the union so why not I?

At that time the initiation fee was 35 cents, so I dug up two bits and a dime and presented myself at the crossroads meeting place on top of a high ridge. The meeting place was picketed but everyone knew me and I was passed through with the admonition, "Watch that goat, son, when you ride him. He is awful."

Miners referred to being initiated as riding the goat and when I entered that circle of men I fully expected to face a goat armed with all his butting paraphernalia, but when the meeting adjourned it dawned upon me that I had been the victim of a practical joke.

Several years elapsed before an inkling of what was meant by unionism began to dawn upon me. All these years I plugged along, paying my dues grudgingly and occasionally attending a meeting until the year 1908. I was then living in Olcott,[4] West Virginia, a mining camp, after being married to the sweetheart of my boyhood days. We rented a coal company house, secured a few pieces of furniture on the installment plan, and started in to be regular and raise a family.

One day a salesman stopped by and offered a book, *Hell before Death* by the Rev. W. S. Harris.[5] That title appealed to me because heretofore all I had heard about the infernal regions had to do with the period after death. So I purchased the book. This book explained many things that I wanted to know such as industrial evils, panics, unemployment, intermittency of employment, and who controlled finance and politics. After reading this book and absorbing its contents, I drifted into the habit of shooting off my lip every time something occurred that did not appear to me to be just exactly right. Someone plastered me with the name "agitator" and it stuck. An agitator is a man or woman to whom conventional environment has become monotonous, to whom the existing order of things holds no emotional appeal. They seek by agitation and education to bring about social change.

During those days it was not considered any breach of conventions to absorb knowledge as long as you retained your ideas and conclusions. But immediately you began to impart your knowledge to others you became an agitator. Solomon said that the seeker after knowledge would always encounter trouble. Solomon must have known that no man or woman can retain knowledge, especially if that knowledge affects the social system.

## Chapter II

I was considered an average miner. I was husky, strong, and energetic. Like most miners I had married when young. My wife was above the average. She could take almost nothing in the form of the meagre food obtained and make it into something edible. Her recipes were often sought by other women in the mining town because of her apparent economy.

I "led the sheet," as was said when one miner did more work than any of the others. I figured close, but at the end of each month after deductions for house rent, smithing, fuel, doctor, etc., I did not appear to get ahead and of course children were born and had to be cared for.

After three years of hard work and self denial I looked at my two children dressed in common gingham and blue denim, then I looked at my wife, who was attired in a dress made from the cheap calico which had been secured at the coal company store.

I said to her, "Lillian, we don't seem to be getting anywhere. Here we are. I have worked and you and I have skimped and tried to save something for a rainy day, and we are right where we started. We don't seem to be getting anywhere; it is just work, work, and nothing to show for it but a little food, very few clothes, and a hovel over our heads."

"Yes," she answered with a sigh of resignation, "it seems hopeless and the children are coming on too, and will need books and decent clothing when they start to school."

"Lillian," I replied, "the first day the mine is idle I am going to visit another mine and see if I can secure a better job. You know I work every minute that there is to work and yet I can not even get enough ahead to move out of here."

"But how will we move?" she asked. "We cannot move on nothing."

"Well, maybe I can find another job where the company will move me or advance the money to move our furniture."

So the first idle day found me at the depot to board the morning train. It was unusual to see me at the depot and of course some inquiries were made. The company officials learned that I had gone away that morning to hunt another job, so when Lillian visited the company store to obtain the day's supply of food, she was informed by the bookkeeper that no more scrip would be issued to me.

Scrip was a form of punch out card, similar to a meal ticket. This scrip was good only for merchandise at the company store. The color mounted to Lillian's brow when she was told that no food could be obtained that day.

"Why?" she said. "Fred has wages due him in the office."

"Yes," said the bookkeeper, "we know that, but in the event he leaves our employ it may be some time before he moves out of our house and the money due him will be absorbed for house rent."

To say Lillian was mad would be putting it mildly. She called the bookkeeper names, such as grafter, cheat, etc.

"My husband has been slaving here for three years," she stormed, "and hasn't even made a decent living for us and now that he has gone out of the creek to look for a better job we are denied food, when he has money coming to him. Oh! I hope the union comes here, then you will pay wages and we will get something besides scrip for Fred's work."

By this time the superintendent had overheard the argument and when Lillian wished for the union he became furious. "I know you will move now," he said, "and what's more you will move right away, because we don't want any union sympathizers here and we are not going to have them."

When I returned home on the evening train I found my wife in tears. A notice was served on me to vacate the coal company premises in three days or be evicted. Lillian tried to explain to me—she thought she had done something to injure my chances. She told of being denied food and of the scene that followed.

I was thoughtful for a time and then I said, "Well, girl, we will get along some way for we cannot become much poorer than we are now."

11

But I loved my wife for what she had said to the coal company officials. She had said things that I had never been courageous enough to say. She had voiced the sentiments that had burned at my vitals for years, but because of fear of the blacklist I had kept silent. Here was my mate, along with others like her, courageous enough to speak her mind. Some of them were to become leaders of men and women.

I am incapable of describing the courage displayed by the heroic women who passed through the strikes that followed. I only make a feeble effort to do so.

So we moved down the creek "a ways," as it was called, chasing that ever fleeting shadow, a job, the most elusive thing known to him who toils for a living, the thing that is owned by him who employs, and to which one cannot attach himself unless a profit will result to the employer because of that connection.

And it will ever be so. Just so long will the "human drift" be composed of workingmen migrating from place to place, from one point of the compass to another, with the thought ever in mind that just beyond is something better than what I have here.

But when the new land of hope is reached and the grill and grind again begins, disillusionment is rapid and merciless. It saps the strength, destroys the ambition, murders ideals.

We found our new location no better than the old. My earnings were about the same, my expenses identical. I became embittered because of my inability to get ahead.

\* \* \* \*

Rumors of the union coming into our camp reached my ears at times. I made half spirited inquiries concerning the union. I heard of the short hours, the better pay, the protective measures practiced by the union against the blacklist and unemployment.

One day I said to my "buddy,"[1] "Joe, we ought to get the union in here and organize. We would get more wages, our working hours would be shortened, and we would have some degree of protection from the whims of the employer."

Joe looked at me aghast. "Do you know what you are saying? We will get discharged, run out of the creek, and maybe beat up if the guards learn of what you have said."

"I don't give a damn," I replied. "I am getting tired of loading 3,000 pounds for a ton of coal, of slaving from year to year with nothing to show for it, my wife and babies almost naked, barely enough to eat, and no chance to get ahead. Look at my wife and yours, they are wasting away, more because of worry than anything else, trying to make both ends meet. My wife has not been to see a circus for years, she never goes to visit her folks, because she is ashamed of her clothing. Are we forever going to stand for such as this?"

Joe sat down and after several moments of meditation, he asked, "Well, what are we going to do about it?"

"Let us talk to the other boys on the Q. T. and see how they feel about the union," I replied.

We approached several miners that day and they, too, were sick of the conditions that prevailed and it was agreed that I should write a letter to union headquarters or go there and see the officers.

One day when the mine was not working I boarded a train up the river to one of the mines that was working under contract with the union. I must have talked to the wrong man at the mine for the company guards had learned of my doings and I was informed after boarding the train that if I visited union headquarters that day that I must not undertake to come back to the creek that evening.

This made me good and mad. I said to the guard, "I will go to union headquarters in spite of you, hell, or anything else."

I visited union headquarters and talked to the president concerning organization at the mines. They were not ready to begin at that time but he advised me to go back and remain quiet until they were ready.

"But they found out that I was coming here and notified me that if I did so to not undertake to enter the creek this evening," I said.

"Now that complicates matters," he replied. "You may have to go up the river and secure a job at one of the mines where the men are organized."

I left union headquarters and returned home unmolested other than to be told when leaving the train to get out as soon as possible, or I would be carried out.

Realizing the significance of that warning, I rented a house from a private citizen and moved into it.

The union mines were along the Kanawha River into which the creeks referred to emptied. To one of these mines[2] I wended my way and asked for a job.

"Where are you from?" asked the job foreman.

"Cabin Creek," I replied.

"Don't know whether I had better hire you or not," he said, "the men here are skeptical about men from Cabin Creek. We have had several gunmen come here and get into the union as spies."

I looked at him askance. "Do you mean to tell me that those thugs come here and go to work in the mines and then join the union?"

"Yes," he replied, "and when they do get in they report the activities of the union to their chief."

I was stunned. This was my first introduction to the "agent provocateur." How important a part the agent provocateur was to play in my life never occurred to me at that time. But a glance backward causes me to wonder at the gall of these sewer rats with serpentine mentalities.

I told the foreman a few things, what I had been up against, how they had ordered me out and what for, and he hired me.

But I did not move to these mines until some time late in the fall of 1912. Martial law was shortly declared[3] and as permission to move had to be secured I did not relish the undertaking.

I would go to my home on Saturday evening and return on Sunday afternoon to the mines where I was employed. On one Saturday evening when I returned home the lieutenant of the militia came to see me.

"You will have to mow the weeds and whitewash the fence around your yard," he ordered.

"I don't have to do anything but die and pay taxes," I said.

"You will cut those weeds and whitewash this fence or I will put you under guard and make you do it," he added.

"Listen," I said, "I am working at Cannelton, do not arrive home until late Saturday evening, and leave in time for the 3:00 p. m. train on Sunday. I haven't time to cut those weeds and whitewash that fence and I do not have the money to pay someone for doing it."

"Well, we will see about it," he said as he walked away.

When I came home on the next Saturday afternoon the weeds were cut and the fence whitewashed. The lieutenant had placed an old bum under arrest and made him do the work. The strikers poked fun at this man because of his being forced to perform a little manual labor at least once in his lifetime.

* * * * *

My first real introduction to unionism came during the conflict of 1912-1913. It was in that year that the great Cabin and Paint Creek strike was called by the miners' union. Working agreements between miners and operators expired on March 31, 1912. The coal operators refused to renew wage agreements.

In the early summer, after every honorable effort had been put forth by the union officers to secure an adjustment of the wage controversy and recognition of the union, a strike order was issued, effective at all mines where coal companies had refused to sign wage agreements.[4]

Cabin Creek had been non-union since 1904,[5] at which time the union had been crushed by gunmen in the employ of the coal barons. These gunmen were unscrupulous and evicted miners from coal company houses without due process of law. They framed and jailed miners who would not agree to their terms. These gunmen cracked heads, maimed, and in many instances killed miners outright because they would not renounce the union.

Therefore, when the strike was called affecting Paint Creek, it spread to Cabin Creek, where miners had been smarting under the lash of the gunman system for eight years. Wholesale evictions followed after the strike call had been nobly responded to. The coal barons imported gunmen by the hundreds. The majority of these gunmen were secured through Baldwin-Felts Detectives Incorporated of Roanoke, Virginia.[6] Even before the strike was inaugurated the coal companies had shipped in many automatic machine guns, hundreds of high powered rifles, and supplies of ammunition.

The meager belongings of the striking miners were piled out along the highways; heads were cracked and in one instance a miner was dispossessed and his wife gave birth to the coming generation while lying on a mattress by the side of the road. A mine guard was charged with kicking this woman while the evictions were taking place, but no conviction could be secured.

Tent colonies began to spring up in every spot where private property not owned or controlled by the coal companies could be secured. Into these tents the evicted miners moved their families and their earthly belongings. It became a pastime for the gunmen to fire on these tent colonies from ambush in the hills. Every day or two they would sneak into the hills and sprinkle the canvas cities with showers of leaden pellets, caring not if their bullets hit men, women, or children.

These miners were the descendants of the pioneers who had hewed their way through the forests of western Virginia, driven the Indians out of the territory, erected log cabins, and cleared farms. In many instances the land from which the miners were driven had been owned by their forefathers until they were robbed of it because of their inability to understand the law.

It was not long until the miners began to retaliate. They purchased obsolete army rifles such as those used in the Spanish-American War. The price of these rifles at that time was about $1.98 each. Many of the men owned muzzle-loading squirrel rifles, but they soon learned that they could not compete with modern, up-to-date rifles like those furnished the gunmen by the coal barons.

While this strike was in progress, riots, as they were termed by the subsidized press, often occurred. In every instance where the miners were fired upon by the gunmen the press ignored the occurrence, but if the miners retaliated by returning a few shots subsidized editors would become frantic in their hue and cry for federal or state intervention in the form of martial law.

Many deportations took place in the territory on strike. Men were led out of the strike zone in the dead of night; some were kicked out, others were prodded out at the ends of rifle barrels. Shooting from ambush was a common occurrence.

* * * * *

At Dry Branch[7] on Cabin Creek just prior to Labor Day, Tom Hines, who was then captain of the gunmen and hated by the miners, attempted to kill a railroad man by the name of Hodge. The gunmen knew that Hodge was friendly to the miners. Hines must have decided to put Hodge out of the way. He arrested Hodge and handcuffed him and led him to his (Hines') house.

The story of Hines' wife and of Hodge was to the effect that Hines started to kill Hodge after they were inside the house. Mrs. Hines intervened and Hodge ran out and took refuge behind an empty oil barrel. Hines finally kicked his wife loose and fired six shots through the barrel behind which Hodge was hiding, but Hodge had the presence of mind to lie flat on the ground and Hines' bullets perforated the barrel too high to do Hodge any injury.

After Hines had emptied his gun, Hodge jumped up, ran to Hines, and disarmed him before he could reload. At the time Hines was taking Hodge into his home the miners were securing their guns and getting into the surrounding hills. When Hodge secured Hines' gun, he started to run and another gunman fired at him as he ran and broke his arm at the elbow. Hodge fell to the ground and lay there. Bullets flew. They thought Hodge was dead and from every headstone in three cemeteries there leaped a spurt of flame. From the hills and from behind trees, bullets sped in the direction of Hines and his men. Hines fell mortally wounded.[8]

Both Hines and Hodge were rushed to the hospital. They were put aboard the evening train and taken into the baggage car. On the way to the hospital the baggage master noticed Hodge spitting tobacco juice at the guard's face. When he looked at his face he saw that the guard's eyes were full of juice. The guard was dying then and Hodge was credited with saying to him, "You'd better die, you son-of-a-bitch, for if you don't, I am going to kill you the minute we get out of the hospital."

Several of the gunmen working under Hines' direction were killed or wounded, and when the state militia arrived next morning they smuggled the bodies of the dead and wounded gunmen out of the mining camp.

On the morning after the fight, the governor arrived with his staff to conduct an investigation and fix responsibility, if possible. But the governor and his staff, which consisted of the officers of the militia, lost sight of the part played by the gunmen in that disturbance and only tried to fix responsibility for the part played by the striking miners. The state was then under a Republican administration. The governor was William E. Glasscock of Morgantown.

When the governor arrived, he approached two miners who were sitting on the porch of the coal company store. "Boys," he said, "what seems to be the trouble around here?"

One of the miners looked across at the hills and said, "Well, you know, Buffalo Bill, Kit Carson, and Texas Jack live over beyond them hills, and they occasionally come over on this side for a little target practice." This answer constituted the information obtained by the governor on that occasion.

Several days after the death of Hines, martial law, with all its perversion of justice, was declared.[9] The coal operators had been frantic in their appeal for martial law. Leading daily newspapers had voiced the sentiments of the coal barons, for the mine owners either owned the newspapers or had subservient tools on the managing and editorial staffs.

When I retired on the night before martial law became effective, there were seven "tin horns," as the militia were called by the miners, standing at the entrance to the front yard.

I remarked to my wife that "All we do, or any place we go tomorrow, will be done only with the permission of the military authorities, and you are going to lose the oil cloth cover off your dining table."

"What are you going to do with my oil cloth?" she inquired.

"Wrap my guns in it and take them to the hillside," was my reply. I sneaked out of bed and took my guns, wrapped them in the oil cloth and climbed out of the rear window, climbed across two fences, waded the creek, and climbed up the hillside to a ledge of rock that I knew about. Pushing them under the outcropping rock as far as they would go, I walled up in front of them with pieces of rock and raked leaves against the rock wall as best I could in the darkness until I thought they were sufficiently hidden from view.

My wife was frightened all the time I was away from the house for the report was out that anyone caught hiding their guns would be sent to the penitentiary.

We did not get up until about 8:00 a. m. the next morning and a short time after the militiamen heard some stirring in the house they knocked loudly on the front door.

When I opened the door they said to me, "As you know, martial law has been declared by the governor and we are instructed to gather up all guns in the zone and hold them until this thing is over. You will be given a receipt for your guns you turn over to us, and when this thing is over you can present the receipt at headquarters and your property will be turned over to you."

"Gentlemen," I said, inviting them into the house, "the only gun here is an old, flip-up block, .32, rim fire Flobert, and it hasn't been fired for so long you can scarcely see through the barrel for rust and dirt. If you wish to take charge of that gun, you are welcome to it."

I went to the closet and secured the Flobert and handed it to them. Taking the receipt they offered, I invited them to look the place over. They looked around through the house and went out the door. As they were going through the gate two neighboring women that lived in the next two houses were standing on the porch nearest to my house and one of them said to the other, "They didn't get his guns. Why he has two or three over there."

One of the militiamen overheard the remark and back they came. They gave my house a complete going over. They looked under mattresses and beds, up the chimney, in the attic, for places where a hole could have been cut in the floor, in closets and every nook and corner where a gun could possibly have been hidden. When they had concluded their search and were ready to leave they cautioned me, "We suppose you know what the penalty is if you are caught with a gun that you have refused to surrender, do you not?"

"Yes, I have heard that it might carry a term in the penitentiary," I replied.

How stringent the rules were to be never occurred to anyone until we undertook to visit the grocery, the post office and other places to which we were accustomed to go. Permits had to be secured to do anything we wanted to do or to go any particular place we desired to go.

A military court was established at Pratt; a bull pen was erected and scores of miners were arrested and imprisoned there. Many of them were given a pseudo trial and sentenced to the state penitentiary at Moundsville, some for terms as long as 20 years. This "drum-head" court, as it was termed by the attorneys representing the miners, was presided over by Judge Advocate George S. Wallace of Huntington.

No mercy was shown the striking miners. They were allowed no defense in that court; in fact it deserves no comparison to any institution other than the Roman Inquisition.

A few test cases were carried to the Supreme Court of the state on appeal and one of the most feudalistic decisions of modern times was rendered in their cases. Hon. Ira E. Robinson, a member of the Supreme Court, wrote a dissenting opinion in these cases, but Judge Robinson aspired to the governorship of the state and many accused him of resorting to political expediency.[10]

Prior to the declaration of martial law and during the first months of the strike a commission was appointed or elected by some authority, I do not remember who. The commission was headed by the Right Rev. Bishop Donahue of Wheeling. This commission was to inquire into the causes of the strike, but the inquiry lost sight of the industrial struggle entirely and spent most of its time chasing the bogy of socialism.[11]

* * * * *

But the "drum-head" court was in for a grilling. Mass meetings were held in the communities outside the martial law zone. Resolutions of protest were adopted and committees were appointed to call upon the governor and ask him to withdraw his proclamation. Along with 34 others, I was appointed on one of these committees to visit the governor.

Among the 35 was Mother Jones, the white haired angel of the miners. Mother Jones was then about 80 years of age. Her hair was snow white, but she was yet full of fight. With that brand of oratorical fire that is only found in those who originate from Erin, she could permeate a group of strikers with more fight than could any living human being. She fired them with enthusiasm, she burned them with criticism, then cried with them because of their abuses. The miners loved, worshipped, and adored her. And well they might, because there was no night too dark, no danger too great for her to face, if in her judgment "her boys" needed her. She called them her boys, she chastised them for their cowardice, she criticized them for their ignorance. She said to them, "Get you some books and go to the shade while you are striking. Sit down and read. Educate yourselves for the coming conflicts."

The committee of 35 on which Mother Jones and I served boarded the train for the statehouse. We left the train at the depot in Charleston and were walking through the city to the governor's office. As we approached the statehouse from the east side, Mother Jones was singled out and arrested by Dan Cunningham and Howard Smith. These two men were alleged participants in the Stanaford Mountain massacre in 1902. The miners were striking at that time and one night they were fired on while they slept. Shots were fired through the walls of the houses in which they lived. Several of them were murdered in their beds. The mattresses on the beds where they slept were saturated with blood. Cunningham and Smith were acting in the capacity of guards at the time and were employed by the coal corporations.

Mother Jones was walking between a miner by the name of Walter Diehl and me when she was "nabbed," thrown into an automobile, and hurried away. Consternation ensued among many of the committee, but the level headed members maneuvered the committee to the headquarters of the mine workers.

People began hurrying through the street; scores of policemen were running towards the statehouse. We were puzzled regarding the cause of all this disturbance.

It was soon learned that a riot call had been sent to police headquarters and the report made that 3,500 men led by Mother Jones were on the way to kill the governor and dynamite the Capitol. In this report units had been multiplied to hundreds and excitement was at fever heat. They imprisoned Mother Jones in the bull pen at Pratt.

She was held incommunicado for some time, not allowed to see any one or send a message, but eventually she did get a message to Senator John W. Kern of Indiana. She often said of him, "He was a man on whom labor could depend when his services were needed."[12]

Upon receipt of the message from Mother Jones, Senator Kern introduced a resolution on the floor of the United States Senate calling for an investigation of the conditions prevalent in the coal fields of West Virginia.

When the resolution was introduced and a sub-committee of the Committee on Education and Labor was given the power to inquire as to whether or not civil government had broken down in West Virginia, all the strikers sentenced by the "drumhead" court set up under martial law were immediately given their liberty. Jails and the penitentiary began to disgorge their prisoners. No explanation was given as to the reason they were released. They were just told that "You can go home."

The Senate committee found that civil government had become subservient to the local coal barons and that wages paid miners and other labor in and around the mines were far below that level which could possibly be classified as a minimal standard of living.

This system was chiefly maintained by coal barons who lived in Millionaire's Circle in Cincinnati, Manhattan and the Bronx in New York City, and in Boston. Absentee landlordism governed the state from magisterial districts to the gubernatorial chair.

From these men of wealth, who ground their employees into profits in order that they might fare sumptuously and clothe their offspring in purple and fine linen, state, county, and municipal officers took their orders. Campaign promises to remove the gunman system were forgotten when public offices were occupied. The clink of ill-gotten shekels, the metallic ring of the thirty pieces of silver reached the political office holders and caused them to forget even their political platforms.

The men who maintained and perpetuated the gunman system were often played up in the press as philanthropists of the first order because they endowed religious and educational institutions with fabulous sums ground from the miners who produced their wealth. Every dollar in each sum total represented anguish, pain, misery, blighted hopes, crushed aspirations, stifled ideals. Viewing their activities from the proper angle, every dollar that they gave represented the groans of overworked and underfed employees, the pitiful supplications of undernourished women, and the heart-rending whines of starving babies.

Many fair-minded citizens of this nation visited these coal fields incognito and saw for themselves. The United States Senate found and condemned this system. Thousands of citizens from the east and north saw and understood. They visited these coal fields and were asked what their business was, how long they intended to stay, and were escorted to their lodging places. They had their baggage ransacked and when questioned, if their answers were not satisfactory, they were told when the next train pulled out and that they had better be aboard it, and that if they were not they would either be given lodging at the county jail or have their heads cracked and be deported out of the community.

The chairman of the committee which investigated conditions in West Virginia was Senator Claude A. Swanson of Virginia. The other members of the committee were Senator John K. Shields, Tennessee; Senator James E. Martine, New Jersey; Senator William S. Kenyon, Iowa; and Senator William E. Borah, Idaho.

This committee inquired into the living conditions, wages paid, hours of labor, educational opportunities, recreational and social activities, etc. Industrial and political conditions at the time of the strike in 1912 bordered on peonage if peonage did not actually exist.[13]

Coal miners were paid low wages; housing conditions were unbelievable unless one observed for himself. Wages for day labor were from $1.85 for common labor to $3.00 for the most skilled. Average gross earnings of miners were from $2.00 to $4.00 per day for each day actually employed.

For example, the employees of the Coalburg Colliery Company, 117 in number for the month of July, 1912, earned $7,436.33, an approximate average of $63.00 per employee. Deductions for house rent, store account, lights, doctors' fees, coal, etc., were $6,118.80, an average of $52.00 per employee. The money paid out on pay day to these employees was $1,880.10, an average of $16.00 per employee, for one month's work.

Miners were maligned, abused, jailed, bull-penned, sentenced to the penitentiary, classified as outlaws and socialists, and threatened with deportation because they demanded more. There were no compensation or other laws for the protection of employees. In the event they were injured or killed their dependents were left to the mercies of the world.

## Chapter III

The mining town at Mucklow was the first operation one encountered upon entering Paint Creek. Paint Creek derived its name from the fact that the water therein periodically changed color. At times the water would show a milky color, then it would change to a purple hue. The Indians named the creek prior to the coming of the settlers.[1]

No private property existed above Mucklow. The coal corporations owned the creek from the first mine to the last, nearly 20 miles of it and from ridge to ridge.

Property on which striking miners could pitch their tents was secured at Holly Grove, several miles below Mucklow. Holly Grove became a canvas city almost overnight. From a place where there was a post office, two or three stores, and only four or five dwellings, it spread out and up and down the creek. Long rows of teepees and wall tents nestled along the banks of Paint Creek and even into the ravines.

The population of Holly Grove ran into hundreds, consisting of striking miners, their wives, and children. The gunmen would often sneak into the surrounding hills, and with high powered rifles they would send hundreds of leaden death messengers into the tent city. Bullets would perforate these tents and men, women, and children would fly for shelter.

But the strikers tired of this practice of the gunmen and they reasoned that if the "thugs," as they were often termed, were given a dose of their own medicine, it would teach them a lesson.

So one evening a powwow was held. About this time one of the largest department stores in a nearby city had shipped several cases of 45-70 Springfield rifles and 10,000 rounds of ammunition to the strikers at Holly Grove. This merchant had heard of the practice of the gunmen and knew the strikers were poorly armed. He shipped the merchandise not knowing whether he would ever receive payment or not.

The powwow resulted in a decision to hit back at the gunmen the next morning. About 3 a. m., 16 strikers met at an appointed place and Indian file they crept and crawled to a vantage point just across the creek from the company store at Mucklow.[a]

They waited for the dawn, that their aim might not be impaired, because they were embittered by the treatment accorded them by the employers through the operations of their hired gunmen. They were not going to waste ammunition shooting for fun. Fresh in their minds was the abuse heaped upon them when the gunmen were evicting them from their homes. Some of the wives of the strikers had been insulted and made to wade the creek while vile language was hurled at them.

If we make brutes of men, can we blame them if like brutes they turn and rend when they gain the upper hand?

When the light began to show in the east, some of these men were hard to restrain. "Wait," said the more wise, "until it gets lighter, so we won't miss any of them."

"I can't wait," said another, gritting his teeth in rage, "one of their bullets struck my wife in the ribs yesterday when the damn things were shooting into the tents."

"Hold on," urged another, "I too have a grudge to settle and I don't want my chances spoiled. My head is still sore from that crack one of them gave me with the butt of his gun."

When it was light enough to distinguish objects at a distance, the gunmen, who were getting ready for breakfast, were surprised by a fusillade at close range.

"I got mine," said one, "he will never mash any more noses."

"Yes, and I knocked mine over," growled another, "he will quit shooting at people while they sleep."

After firing several scores of shots they retired as quietly as they had come, leaving dead and wounded gunmen behind them. How many they never knew; the undertaker who embalmed the bodies and the trainmen who brought them out said there were 13.

✧   ✧   ✧   ✧   ✧

A scene of much action during the strike was Eskdale, a typical mining town situated in the center of the Cabin Creek coal field. The only beautiful thing about Eskdale was its name. It was smoky, sooty, and grimy. The constant puff of railway locomotives, the crash and grind of engines, cars, and trains, night and day, produced an atmosphere similar to that surrounding one of the great steel centers like Homestead, Youngstown, or Pittsburgh.

Eskdale was incorporated. It had a mayor, council, and one policeman. When the coal strike became effective in Paint Creek and the lower extremities of Cabin Creek, the miners working in the mines above Eskdale became restless. Miners were assaulted and driven out by the gunmen, chief among whom was one Don Slater. Slater was a bruiser. He was bold, ruthless, and entertained no scruples against taking human life. He passed up no opportunity to crack the head of a striker or even a miner who dared to talk of unionism.

Eskdale became a rendezvous to which the strikers came for refuge when they were beaten up or told to get out.

Union headquarters were visited time without number and no one could be induced to go into the Cabin Creek coal field beyond Eskdale.

One day a miner named Frank Keeney visited union headquarters and asked for someone to accompany him into Cabin Creek and organize the miners. He met with a blunt refusal. So he proceeded to read the riot act to the union officers, and said to them, "I will find someone with nerve enough to go with me, for if you men are afraid to make the trip, there is a woman who will go."

He proceeded to locate Mother Jones and after a thorough understanding was reached, a date was set for Mother Jones to go into the forbidden territory.

I was standing on the bridge at Cabin Creek Junction the day Mother Jones entered Cabin Creek.

Her hair was snow white, but she could walk mile after mile and never show fatigue. When we saw her drive by in a horse drawn vehicle we knew the meaning of that visit and we fully expected to hear of her being killed by the gunmen. She arrived at Eskdale without mishap, but after she passed through the business center of town and as she approached the southern residence section a body of gunmen could be seen just ahead.

The morning sun cast its rays on the steel of machine guns, behind which stood creatures that could have been men, and as Mother Jones came near the frowning muzzles of these death dealing implements of war, some of these gunmen fingered the triggers of the guns and licked their lips as though thirsty to shed human blood.

But she drove her rig near and one of the miners assisted her to alight. She surveyed the scene with a critical eye and walked straight up to the muzzle of one of the machine guns and patting the muzzle of the gun, said to the gunman behind it, "Listen here, you, you fire one shot here today and there are 800 men in those hills (pointing to the almost inaccessible hills to the east) who will not leave one of your gang alive."

The bluff worked; the gunmen ground their teeth in rage. Mother Jones informed me afterwards that if there was one man in those hills she knew nothing of it. Her comment in regard to this day was, in part, as follows:

"I realized that we were up against it, and something had to be done to save the lives of these poor wretches, so I pulled the dramatic stuff on them thugs. Oh! how they shook in their boots, and while they were shaking in their boots I held my meeting and organized the miners who had congregated to hear me."

The miners took courage after this meeting and many other meetings were held. The coal industry in Cabin Creek became paralyzed. The gunmen swore vengeance against Mother Jones. On every hand threats could be heard. The coal barons appealed to the governor to deport her. She was asked to leave the state, but they did not know Mother Jones.[3]

Eskdale became the center of activity; it was a haven of rest for the embattled and oppressed miners.

The services of railroad engineers, firemen, and conductors were enlisted. Literature, handbills, etc., were distributed at night, scattered everywhere. When the guards got into action each morning they found miners reading handbills that had been distributed during the previous night.

* * * * *

Eventually Eskdale, like Holly Grove, became a thorn in the flesh. The coal barons tried to purchase the town from the property owners. They were told it was not for sale.

Firing a hailstorm of bullets into the little town became a hobby with the gunmen. There were several vantage points from which machine guns could be trained on the town, and frequently showers of leaden hail would sprinkle the tent colony and surrounding buildings.

One day two miners were talking about the danger to which the women and children were subjected. One of these miners was nicknamed Bullethead, the other was referred to as Bad Eye. Said Bad Eye to Bullethead, "Let's stop this business of shooting into the tents. Let's turn the tables on them and pepper them awhile."

"All right," agreed Bullethead. "We'll get a few men together and give these thugs a dose of their own medicine."

Prior to this time a club of the American Rifle Association had been organized at Eskdale and many of the miners had joined. Through this club and to its members Krag Jorgenson rifles were sold for a very small price. There were dozens of these rifles in Cabin Creek when the strike occurred.

In selecting the men to entertain the gunmen and stop them firing on the tent colony, Bullethead and Bad Eye decided on Few Clothes as one of the squad.

Few Clothes tipped the scales at 252. He was encumbered with little if any superfluous flesh; his arms were long, and at the end of each arm hung a fist that resembled one of Armour's picnic hams. It was rumored that Few Clothes was a member of the U. S. Cavalry (colored) that at one time undertook to shoot Brownsville, Texas, off the map. Anyway, he knew how to use a gun.

Only a handful of men were selected to annoy the gunmen, ten or eleven to be approximate, and these men became known as the dirty eleven. Gunmen were kept too busy after the eleven became active to do much shooting into the colony.

The fighting in Cabin Creek was chiefly confined to skirmishes, bushwhacking, and shooting affrays from ambush between strikers and gunmen.

It was considered a part of the duty of strikers to meet transportations of strikebreakers at Cabin Creek and turn them back, if possible.

These transportations were brought in from the cotton fields and industrial centers of the South. They were recruited from the states of Georgia, Alabama, and the Carolinas, both colored and white. The majority of them were illiterate and knew nothing at all about coal mining. They had no conception of what they were running into. They had been told stories that made them believe that money grew on trees around the coal mines and fortunes were theirs for the picking.

I often assisted the men who were assigned to check this influx of human beings, and efforts to talk to them often resulted in a fight.

Few Clothes and a striker by the name of Charlie Stomp and I were at Cabin Creek Junction one evening when about 50 strikebreakers, who were reported to have come from North Carolina, were unloaded from the C. & O. main line train. When they alighted to transfer to the short line train going up Cabin Creek, we immediately approached them. We talked to them and pointed out to them just what they were doing as we saw it.

A tall North Carolinian who seemed to be a leader said to us, "Stand aside there and let us pass, we've no time to 'argy' with you."

"Yes," I replied, "you should give us time to argue with you because you are going up Cabin Creek to take our jobs and it isn't fair to do it."

"Out of my way," he rejoined, as he swung wildly at Stomp.

If that blow had landed as it was intended, Stomp would have gone down for the count. But Stomp was good on the dodge. He was always a good dodger. He could squirm out of most anything or any position no matter how serious or embarrassing it might be.

The blow of the "Tar Heel," as the Carolinians were called, spent its momentum on the air; before he could regain his balance Stomp landed a right swing on his jaw. The blow should have felled an ox, but not the "Tar Heel." He only shook his head and gathered himself together for another swing.

But by this time Few Clothes was in action. "Let me at him, Missa Charlie," the Negro said, as he braced himself against the steel pipes that encircled the yard. Few Clothes threw his 250 pounds of weight into that blow. He landed squarely on the point of the North Carolinian's jaw.

It seemed as if he was trying to fly. He hovered in the atmosphere for an instant and after striking the ground he slid along for several feet and lay still. The fight became general; blows could be heard on all sides. Remarks were exchanged such as "You dirty scabs, you will come in here and take our jobs, will you?" "Take that you damn 'Red Neck',"[4] one striker was heard to blurt as he landed a well directed punch.

Several railroad men joined in this melee and the fight lasted only a few moments. When a strikebreaker would get a real punch, he would get up and run for the main line train and get aboard. The fists of Few Clothes could be seen swinging this way and that. He seemed to be at home here. Five and six men would be pummeling him at one time and the only apparent effect of their blows was an occasional grunt from the big Negro. When one of his blows landed, a man would go down for the count.

When the fight was over, all the strikebreakers had reboarded the main line train (as they could not reach the short line for we had them cut off) with the exception of the tall North Carolinian. We picked him up and placed him aboard with his friends. He had begun to show signs of returning consciousness.

This was only one of many such occurrences, when the strikers would try to check transportations of strikebreakers.

✢ ✢ ✢ ✢ ✢

About the middle of the summer of 1912, rumors were whispered about concerning a concentration of the gunmen at Mucklow on Paint Creek. Mucklow was spied upon by the scouts of the strikers. It appeared that reinforcements had been obtained, and that there were 30 to 40 gunmen at Mucklow. It was learned that they contemplated another onslaught on Holly Grove. The strikers decided to attack first. Runners were sent out and help enlisted from the mines along the Kanawha River.

By this time there were squads of minutemen in each locality, only a few at each mine, ready to go to the relief of the strikers when they were in danger. The hour of the day or night when they were called upon was of no consequence.

All that was necessary to obtain the help of these minutemen was to get word to them and they were off. The minutemen were notified by runners; then they would try to enlist new recruits. Holly Grove was in danger. That was enough. Freight trains were boarded by the minutemen, and one or two would work their way over the train to the engineer.

"Will you stop and let us get off at Paint Creek?" they would inquire.

"Boys, I have no orders to stop, but we usually stop along there to raise steam so when we run out of steam you can get off." The trains invariably ran out of steam just across the river from the mouth of Paint Creek.

One night, men were mustered to reinforce the strikers at Holly Grove; about 100 men boarded the first freight train headed in the direction of Paint Creek. These men were native mountaineers. Their forefathers had migrated into those hills several generations ago. They were determined and unafraid. When they left the freight train opposite Paint Creek, boats were waiting to ferry them across the river.

We were told when the Paint Creek side of the river was reached that Gaujot (pronounced Gasho) had been killed that evening while firing on the tent colony at Holly Grove.

Gaujot was of French descent. He was about six feet four or five inches in height and weighed about 200 pounds. His head was of a peculiar shape; on top of what should have been his head, there was an extra knot or bump about the size of half a cocoanut. Gaujot was feared by the strikers; his reputation was that of a killer.

When word of his death reached the head of the column it was passed along as follows: "Good news, boys, Gaujot is dead."

"Damn, that makes me feel good," said another.

"Wow!" blurted the third. "I ain't been so happy since Pa died."

"Just another rat gone," said the fourth, "hope now soon the rest of them will get it."

But when we reached Holly Grove, we learned that it was not Gaujot but Stringer that had been killed. All of us walked up the track to see the body of Stringer. It was told afterwards that Stringer had imbibed freely of fire water and being all bad anyway, he had challenged the other gunmen for any one of them to ride through Holly Grove on a velocipede and shoot up the tent colony.

"You are a gang of cowards," said Stringer to the other gunmen. "You haven't got gall enough to face these 'red necks.'"

"It is like committing suicide," rejoined another gunman named Phaupp. "but if nothing else will satisfy you, come on, I'll go with you."

They boarded the velocipede and started for Holly Grove. As they approached the tent colony Stringer pulled his pistols from the holsters and began shooting. The strikers were expecting trouble and pickets were out, and when Stringer began shooting they called to him to halt. He answered with another volley and from several points he and Phaupp were met with rifle fire. Both men fell to the ground, Stringer dead, Phaupp seriously wounded.[5]

I looked at what had once been Stringer. The hole through his head appeared as though a pool ball had gone in at one side and out at the other. His body was not moved until the next day. Phaupp crawled away and reached the Sheltering Arms Hospital at Hansford, at which place his wounds were dressed.

The night was spent resting on guns, every man ready for a squall. Scouts were going and coming all night. At 3:00 a. m. the next morning a body of approximately 300 men moved quietly out of Holly Grove and after silently wading the waters of Paint Creek crept into the mountains.

Mucklow was to be attacked from both sides of the creek. No conversation was indulged in, other than whispers. Points of vantage were reached, such as old mine openings, tram road grades, holes where trees had been uprooted, etc. Daylight was waited for, the signal to begin firing was the call of the bobwhite. Daylight appeared; men began to stir in the mining town, gunmen walked away from the club house to take up their various stations, yet no signal to attack. Nervous fingers gripped triggers of Mauser, Krag Jorgenson, and Springfield. These men were mad, they wanted revenge, they were going to get it and delay enraged them.

The shrill call of the bobwhite rent the morning air and before its echoes had reached the opposite hill hundreds of rifles belched forth pent up revenge. Men could be seen below running to and fro trying to gain vantage points. Machine guns opened up from below and rat, tat, tat and whine of bullets answered the volley of the strikers. Bullets from the machine guns tore into the ground, plowed their way along logs, knocked bark from trees. Two of these of the automatic type were being used by the gunmen. A bullet clipped the hat from one man's head. "Damn, that was a close shave," he said. A bullet struck the stock of another striker's gun and it was split in two. Six mine cars loaded with coal were sitting at the mouth of the mine; these were pushed to the knuckle and turned loose down the plane. They tore their way down hill on the steel rails to the tipple nearly a thousand feet below and wrought havoc in their mad flight until they ended up in a cloud of dust and grime in the machinery of the tipple.

One machine gun was among some piles of railroad ties in the valley, another was in an empty box car on the siding. Several of the best shots were designated to silence the machine guns, if possible. But this was no easy task for as soon as one operative would be wounded or run away from his post another would take his place. It was estimated that 10,000 shots were fired by both sides. It was never known how many gunmen were killed, or if there were any, but reports from railroad men and undertakers set the number at 15 or 16. The strikers suffered no losses whatever.

When the fight was over the strikers scattered into the hills and wended their way home to await the consequences.

\* \* \* \* \*

Holly Grove remained a sore spot to the coal barons, for the non-union miners working in the mines were approached at every turn on the road where strikers could reach them. Fist fights were a common occurrence between strikers and strikebreakers.

It was finally decided by the coal operators and the sheriff of the county that Holly Grove must be blotted out.[6]

The Chesapeake and Ohio Railroad shops were located at Huntington. In these shops a passenger coach was armored with a double sheet of car steel from the floor to the bottom of the windows. Machine guns of the latest type were mounted in this coach, and the crack riflemen of the coal companies were put aboard. The sheriff of the county in which the strike was in progress was also an occupant of the armored coach.

One of the big coal operators cast his lot with those who boarded the death coach, and this "Bull Moose Special," as it was afterwards termed, stealthily made its way from Huntington to Pratt.

Pratt was the junction of the Paint Creek branch and the main line of the railroad. Late at night the death coach crawled through the switch at Pratt and glided slowly towards Holly Grove.

The night was dark and the death train was darker. There was a cut through which the train must pass as it approached Holly Grove; no lights were used on the engine.

It looked the black monster it really was, creeping through the cut, the engineer at the throttle and those would-be killers in the armored coach. Guns were oiled and loaded, fingers twitched nervously at triggers, thirsty to shed human blood.

It was the intention of those who prepared the death train to take the strikers unawares and slaughter them as they slept. Women and children would have suffered the same as men.

But the residents of the tent city had been tipped off and were on the lookout for the armored train. However, they did not expect the train to arrive until after midnight. A large mess tent had been set up in the center of the colony; cooking was done community style and all dined together. To right angles from this mess tent and into the side of the hill a dugout had been made, into which the women and children would retreat when the gunmen would fire on the tent city.

Musical instruments had been brought into the mess tent, and a group of strikers were enjoying violin, guitar, and banjo when the death train crawled through the cut into Holly Grove.

The violinist was making a stroke with his bow when an ominous toot was heard. His arm became as if paralyzed but only for a fraction of a second. He dropped the violin as he cried, "Boys, that's the special."

Instruments were thrown aside and rifles were secured. From music, the most inspiring thing known to man, these men leaped to arms, ready to defend the only place they could call home. From music to death. Can one imagine more of a change from one extreme to the other?

The toot of the engine whistle was a signal to the occupants of the armored coach to open fire. Machine guns spouted streams of bullets from windows on both sides of the coach. Expert riflemen discharged shot after shot into the tents, the train scarcely moved. They were taking plenty of time to execute the strikers.

The fight was not to be one-sided for long, however. All at once spurts of flame leaped out of the darkness on every side. From behind trees, creek banks, and knolls, rifles cracked and bullets shattered glass in the armored coach. A bullet found its way to the hand of the engineer; some of his fingers were shot off. The armored coach was riddled with bullets. It looked like a sieve.

Cesco Estep was one of the strikers. He and his wife lived in a tent near the dugout. Mrs. Estep was outside the dugout when the armored train opened fire. Cesco said to her, "Hurry, dear, get inside quick."

These were the last words Cesco Estep spoke on earth. His wife was no more than inside the entrance when an explosive bullet from the armored coach struck him on the right side of his face. He fell at the entrance of the dugout, his entire face shot away.

Mrs. Estep heard him fall; she walked into the open and stumbled over his prostrate body. When she became aware that he was dead she did not scream as most women do. She felt around until her hand came into contact with the rifle her man had borne. She secured his cartridge belt and loaded the rifle.

Taking her place among the men who were fighting for what they considered liberty, home, country, for the right to live as all men should be allowed to live, she cried out to them: "Make every shot tell boys, do not waste your ammunition. My man is dead; his head is shot off. Give them hell."

Bullets flew! The strikers began anew, and Mrs. Estep was in the thickest of it. The Mauser she had picked from beside the body of her dead mate cracked incessantly, and she was no mean shot.

One striker had his gun shot in two while he held onto it; his hands were numb for hours. The fight became too hot for the death special and soon after the engineer's fingers were shot off he steamed up and pulled out.

The "Bull Moose Special" referred to is a matter of history. But to the sepulchre of Cesco Estep for years afterward, on Decoration Day, a solemn procession of men, women, and children wended their way to place flowery tributes in honor of their beloved dead.

The residents of Holly Grove slept but little that night. Some one said, "Boys, let us get ready, for that train will come back sometime later in the night."

A tilting barricade was arranged in the cut above the tent city, in such a way that it could be tipped over when the train approached. Men were placed at vantage points at the top of the cut on both sides of the railroad. "Boys," said one, "we will clean them up if they come back tonight. They can never get through here now if they try."

But dawn crept into the valley, revealing the ravages of the machine gun fire indulged in by those of the special the night before.

Roll was called and all answered but Cesco Estep. From the tent in which rested the last remains of Cesco Estep, the soft weeping of a woman could be heard. For she had lost her world.

The wounded were few and the wounds were minor, but some of the tents were irreparable. It seemed a miracle that so many shots could be fired and only one man killed. Providence may have intervened, who shall say?

Holly Grove assumed the regular routine again, repaired damage, cleaned and oiled rifles, checked up on ammunition and supplies.

They ate, they sang songs, enjoyed music and started in to take life easy again as strikers are wont to do.

\* \* \* \* \*

This poem, "When the Leaves Come Out," was composed by Ralph Chaplin during the Paint Creek strike in 1913:

> The hills are very bare and cold and lonely;
> I wonder what the future months will bring.
> The strike is on—our strength would win, if only—
> Oh Buddy, how I'm longing for the spring.
>
> They've got us down—their martial lines enfold us;
> They've thrown us out to feel the winter's sting,
> And, yet, by God, those curs can never hold us,
> Nor could the dogs of hell do such a thing.
>
> It isn't just to see the hills beside me
> Grow fresh and green with every growing thing;
> I only want the leaves to come and hide me,
> To cover up my vengeful wandering.
>
> I will not watch the floating clouds that hover
> Above the birds that warble on the wing;
> I want to use this gun from under cover—
> Oh Buddy, how I'm longing for the spring.
>
> You see them there, below, the damned scab herders!
> Those puppets on the greedy owners' string;
> We'll make them pay for all their dirty murders—
> We'll show them how a starveling's hate can sting!
>
> They riddled us with volley after volley;
> We heard their speeding bullets zip and ring,
> But soon we'll make them suffer for their folly—
> Oh Buddy, how I'm longing for the spring.

## Chapter IV

By mid-1913, the Cabin Creek operators had signed working agreements. After the strike was settled peace and quiet reigned where industrial war, bitterness, and bloodshed had left many scars. The miners did not secure all their demands when a settlement was agreed upon, but competitive bargaining is always a compromise and they did secure the right to belong to the union of their craft, along with certain working conditions, such as a shorter workday, semi-monthly pay day, and a grievance committee.

Happiness and contentment prevailed to a marked degree in the mining region and old wounds and sore spots gradually healed until the month of September when the Senate investigating committee held its hearings. The coal operators who had signed working agreements testified that harmonious relations existed between them and their employees.

Membership in the union increased by leaps and bounds after the strike was settled, despite the fact that some of the officers of the union had lost the confidence of the membership and were openly charged as being drunkards and crooks.

In the spring of 1914, a district convention of the miners' union was called to meet in Charleston. Each subordinate local union was entitled to send one or more delegates, according to its membership.

I was selected as a delegate to represent Local Union No. 2900, Cannelton.

I had served as mine committeeman, dues collector, alternate check-weighman, and financial secretary in a local capacity. I had bitterly opposed the terms of settlement agreed to in the Cabin Creek strike and had criticized officers of the union for what seemed to me to be a partial surrender of the miners' demands. For this attitude I was made the victim of calumny by the official organ of the miners' union in West Virginia, the *Miners' Herald*, published at Montgomery. They editorially called me red, socialistic, one of Big Bill Haywood's[1] followers, "the mouther," etc. Immediately after these editorial comments were hurled at me I started after the paper. This mouthpiece of officialdom did not only hurl its vindictive epithets at my head, but it started in to calumniate every leader that had been active in the strike.

I visited many local unions and made what I thought were speeches imploring the miners to refrain from reading the paper until they ceased their campaign of vilification.

The imported editor of this newspaper was deaf. I classified him as being "physically deaf, mentally dumb, and morally blind." The word "Bolshevist" had not yet emanated from dark Russia or the deaf editor would probably have used that toward us during his tirade.

During the time I was fighting this newspaper, I visited the miners' local at Donwood. Their meetings were held in a school house adjacent to the mine. I was shadowed by someone on this occasion as I had been on others and when one of the officers left the school house to make inquiry concerning me, he butted into my shadow. The shadow came out second best in the encounter which ensued. The officer came back all excited. He stopped me from speaking and asked, "Did you know that you were followed here?"

"No, I did not," I replied, "but since I have been fighting that rotten sheet down at Montgomery someone has been trailing me constantly."

"The guy that trailed you tonight will not do so again soon," he asserted, "for he is headed for the surgeon now and probably the hospital."

I assured them I was sorry to have been the cause of any disturbance. "That's all right, boy," he said, "we will escort you to the edge of the town when you leave and they better lay off."

The officers of the union were peeved at me because of my recent activities and when I arrived at the convention I did not wait for them to start the fight. I started it myself.

It had always been customary for the president and chairman to appoint all committees to act in conducting the conventions. I was determined that this was one convention in which the chairman was not going to appoint the scale committee. When the committee on rules and order submitted its report, it provided for the appointment of the scale committee by the chair. I offered the following motion:

"Mr. Chairman, I move you that that part of the committee's report which refers to the appointment of the scale committee be amended to read 'the scale committee shall be nominated and elected from the floor of the convention by the votes of the delegates present!' "

I was accused of trying to obstruct the proceedings. The motion was argued pro and con, but my amendment carried. The personnel of that scale committee were different from any that had ever participated in a wage conference in West Virginia. We negotiated for several days and arrived at no definite conclusions whatever. Some of the delegates would go to their homes in the evening and return to the convention the following morning. One morning a delegate by the name of Lusk called me aside, shoved a letter under my nose, and said, "Look at this, no wonder we can't make a contract."

It had all been agreed to before the delegates were called into convention. I read the letter. "Where did you get this letter?" I inquired.

"One of brother Charlie's kids was playing around the store at Longacre and the waste basket from the company office had been emptied in the back yard and its contents set on fire. A puff of wind scattered the papers and the kid picked this one up and was playing with it because of that colored heading when its mother noticed the letter. She called Charlie up and told him to come home at once and this is what for," he said.

"Get your brother, Keeney, Lindville, and Boswell and we will meet at Daddy Gherkins' house at noon," I suggested.

We met at the appointed place and hour. That letter was analyzed. I was selected to make the fight on the convention floor.

The letter was written by a Mr. Knox of the Kanawha & Hocking Coal Company, Columbus, Ohio, to one S. C. Gailey, general manager of the company's mines at Longacre. In this letter Mr. Knox stated in part: "Had a talk with Cairns and Haggerty. They think a suspension can be avoided. Have complete understanding with them and know about what to expect. Of course, a convention will be called and a scale committee appointed, after which the convention will adjourn. Expect smooth sailing if the radicals can be controlled. I send this in longhand because I do not care to take chances on leaks."

Haggerty was dean of the international executive board of the miners' union at that time and was in charge of the West Virginia situation. Thomas Cairns was president of District No. 17 of the miners' union and chairman of the convention.

It was ascertained afterwards that Haggerty was a stockholder in a mine at Perryville, and when the miners' international union learned of this he was suspended from office.

When the convention again convened I asked for the floor on a point of personal privilege.

After reading the letter to the delegates and explaining how I had come in possession of it, I offered a motion that "a committee be elected from the floor to investigate the subject matter contained in that letter and that the president vacate the chair pending the report and recommendations of the committee."

Pandemonium broke loose. I was assailed and defended from all sides. My motion was bitterly opposed by several of the district and international officers. One delegate presumed to assure the convention "that they had good officers and why attack them when we were trying to make a wage agreement."

My reply to him was, "We are not making any wage agreement. Here we have been for days and days butting our heads against a stone wall, thinking that we who have been selected to do so are negotiating a wage agreement when all the time a perfect and complete understanding exists between our officers and the coal operators concerning the points of controversy involved. Mere puppets, figureheads, we are fighting a sham battle of words, wasting time, depleting the treasuries of our local unions for expenses, while we argue over something which has already been agreed to behind closed doors. Are we going to stand for such procedure as this?"

Cries of "No! No!" went up from scores of throats. The part of my motion which called for a committee to investigate the subject matter of the letter carried, but the resolution to force the chairman to vacate the chair pending the report of the committee lost. Delegate Frank Keeney, B. F. Morris, international executive board member from District No. 17, and I were appointed on the committee.

Mr. Knox was at the Hotel Ruffner. We arranged a conference with him but his replies to our questions were evasive. He did not deny writing the letter but he did deny quoting Cairns and Haggerty. He asserted that he wrote the letter from a general knowledge of the situation and not because of any understanding which existed between Cairns, Haggerty, and himself. Keeney and I were able to discern that Mr. Knox had been warned of our coming and posted concerning the investigation. The committee excused themselves and retired to Bernarding's saloon and over a stein we argued pro and con.

"Who warned Mr. Knox that we were coming?" I asked of Morris.

"I did," he replied.

"You are a hell of a committeeman," blurted Keeney. "That's the reason we could get nothing out of Knox. You had warned him to be careful of what he said."

Keeney and I reported our findings to the convention, and Morris made a minority report. Charges were preferred against Cairns, but he was "whitewashed" by the executive board.

A few days after that the morning papers carried the story of Mother Jones' incarceration in Colorado. The papers stated that the jails being full the authorities had secured permission from the keepers to place Mother Jones in the basement of an old sisters' home, a Catholic institution. I submitted a resolution to that convention as follows:

"RESOLVED, that we criticize the Catholic Church for not protesting against their institution being turned into a military bastile and Mother Jones being incarcerated therein."

This resolution was debated heatedly. Miles Daugherty (now deceased), an international organizer from Pennsylvania, declared that "delegate Mooney has insulted this entire convention and I demand that he apologize."

"I apologize to no man or set of men when I think I am right," I retorted. Daugherty started across the hall after me. He was intercepted by Frank Keeney just in time because I was fighting mad and had secured a chair to use as a weapon. Someone had warned me that Daugherty was an amateur pugilist of no mean ability and I meant to take no chances.

My resolution was tabled and I did not again undertake to have it brought up for consideration.

The scale committee negotiated with the coal operators for three weeks and were no nearer a settlement of the controversy than when we began.

We adjourned sine die and the officers were instructed to issue a strike call effective April 1, 1914. We wanted the check-off for union dues like that then in effect in Illinois, Ohio, Indiana, and part of Pennsylvania. We were on strike the entire month of April. Our grievance was carried up to William B. Wilson, then Secretary of Labor at Washington. He gave us a limited check-off in his decision.

When the delegates were called to re-convene the convention, I was left out, "canned" as they termed it, and another was elected to attend. But the rottenness of the officers was so pronounced that the convention adjourned with the officers under fire and subject to catcalls and jeers.

\* \* \* \* \*

Immediately preceding that convention, charges were preferred against W. F. Ray, C. F. Keeney, Walter Diehl, myself and others. C. C. Griffith, then vice-president under the Thomas Cairns administration, came to Cannelton to prosecute the charges against me. But his appearance resembled the stirring of a hornet's nest, and he was eventually persuaded to tear up the charges and throw them into the Kanawha River from the bridge at Montgomery.

On Friday preceding the date set for the trial of W. F. Ray on Monday, a mine committeeman had stabbed me in the left lung with a knife.

Monday morning Ray wired me to appear at Black Betsy that evening in his defense. I could scarcely stand without assistance and my wife remonstrated with me for going, but after the trip on the train and about two hours rest following my arrival at Black Betsy, I felt equal to the occasion. The trial was set for 7:30 p. m., and Thomas Cairns, Charles Tittle (an appointee), and P. F. Gatens, then international board member, were present. When Cairns was appraised of the fact that I was there to defend Ray, he called his aides together and departed, leaving P. F. Gatens in charge of the prosecution.

P. F. Gatens, Keeney, Ray, and myself had always remained friendly despite the internal strife and turmoil with which the district was infested.

Ray was charged with willfully and maliciously maligning an officer during a debate on the floor of the previous convention.

The records of the convention had never been transcribed so my position was that in the absence of the record, no verdict could be found. In this position I was sustained. The records were never transcribed insofar as I know and no more was heard of the trial.

\* \* \* \* \*

From that time on the officers lost prestige and a group of the union men in Cabin Creek pulled out and established District No. 30. District No. 30 was in for a drubbing. While it broke away from officers who permitted grievances to be settled with the check-book and made working agreements behind closed doors, it had no financing, no backing, and the miners' international union refused to recognize it as a district.

The secessionists selected Frank Keeney of Eskdale as their president, Lawrence Dwyer, as secretary, and Charles Lusk, Sr., as international board member. Lusk did not stick for long, but Keeney and Dwyer pushed on as their colleagues fell by the wayside. Barefooted, hungry, and almost naked, they fought for recruits for the secessionists and in a measure succeeded.

I was in sympathy with the principle being fought for by the secessionists but did not believe that a man or a group of men can put their house or houses in order by going outside and throwing bricks through the windows.

Lawrence Dwyer was often referred to as the walking secretary; his office and equipment consisted of a briefcase wherein a few papers were kept. He was a typical Irishman, capable of speaking in public as most Irishmen are.

Dwyer had lost a limb. He walked around on a peg, yet it was remarkable the distances he was able to cover each day.

Keeney was all fire and dynamite. He asked for and showed no quarter. The international union was eventually forced to assume the autonomy of the district and let the officers out.

George Hargrove and C. H. Batley, two international organizers, were placed in as president and secretary, respectively. The secessionists were admitted under this arrangement to full membership with all rights and privileges as members of the union. The membership increased rapidly under the leadership of these two men and all parties to the controversy were satisfied with the exception of the ousted officers.[2]

* * * * *

In 1915, I was blacklisted for six months. I could not secure work anywhere. I walked for miles and miles up and down creeks and across mountains. Friends would pass my gate and say: "Old boy, how is the food; is it getting low?"

"Pretty low," I would reply. "Split with you directly." Pretty soon some member of the friend's family would come in with a bag of flour, potatoes or other provisions, but at times while on the road I went hungry.

The coal company where I lived and was employed closed down a part of the mine in order to do me out of a job, but work was never resumed on the job where I had been employed until I was given my working place. Dozens of men were asked to take my place. "No!" they would say, "you shut that part of the mine down to get rid of Mooney and that coal can lay up there until Gabriel blows his trumpet and we will never touch it."

One day the manager sent for me and said: "Mooney, we have decided to put you back to work."

"For your sake and mine, I am glad that you have, sir," I replied.

"What do you mean?" he inquired.

"I meant just what I said, sir. If you recall, you told me once, 'Mooney, I will starve you into submission one of these days,' and I asked you how you would go about it, and you said 'I will maneuver you out of work and prevent your obtaining employment until you do submit.' Do you remember what I said to you on that occasion?"

"No, I don't believe I do," he replied, "but we are going to put you back to work. You can call at the office for checks now."

I thanked him and retired. It was near the holidays. Christmas was approaching and the day before Christmas I was standing in the coal company store feeling blue. This would be the first Christmas since I was married in 1908 that toys, candy, and food were not in the house.

An old fiddler by the name of Smith came into the store.

"What are you looking so blue about?" he inquired. "Have you gone to work yet?"

"Yes," I replied, "they put me to work last Monday."

"Here," he said, as he thrust $15.00 into my hand, "get you some Christmas and forget it."

To say I was touched would be putting it mildly. I blubbered a "thank you" but he was gone out through the door. Afterwards, when I offered to repay him, he declared: "Why man, you are crazy. I never loaned you a penny in my life."

During the time I was blacklisted another miner killed a large hog and divided it in half, threw one half into a wagon, and brought it to me.

"I will pay you for this some day," I asserted.

"You ever try it and you will lose a friend," he said.

This is the spirit of fellowship, love, and devotion that permeates the life of the union coal miner. He will give until it hurts and then divide what is left with his fellow men.

✦ ✦ ✦ ✦ ✦

In 1915, while living at Cannelton, I signed up with the American Correspondence School of Law for a law course. I received all the textbooks and went along fine with the lessons for some time until along with one lesson came a lecture on private property by Charles E. Markus, L. L. D., Ph. D., etc. To absorb and believe this lecture, private property was the God of everything living or dead. It infuriated me to the extent that I quit the course and sold the books.

I delved into the works of Victor Hugo, Engels, interpretations of Marx by Engels, Marcy, and others. The books of Jack London, Voltaire, Plato, Balzac, Thomas Carlyle, and Herbert Spencer were also included. Marx himself I could not absorb. His work was too difficult for I had only reached the sixth reader in public school. I never studied algebra, trigonometry, and geometry until I was 30 years of age.

Many times during the years I worked in the mines I did without a shirt or some other item of clothing in order to be able to purchase a book I wanted to read. Never will I forget when the news leaked out at Cannelton that I had sent for the *Origin of Species* and the *Descent of Man* by Darwin. The store manager's wife learned of it and one day she cornered me in the store. "Mr. Mooney, they tell me you are going to read Darwin?"

"Yes, Mrs. Scott, I am," I replied.

"Mr. Mooney, you cannot afford to read that stuff," she pleaded.

"Why not? Mrs. Scott, I don't have to believe what Darwin wrote," I said to her.

"No, but you cannot read that kind of literature without absorbing some of it," she admonished.

"Yes, I believe I can, Mrs. Scott," I assured her.

She finally gave me up as hopeless and I read Darwin through. I did not and do not yet believe half of what he wrote.

During the year 1915, someone sent me a copy of the *Menace*, an anti-Catholic newspaper published in Aurora, Missouri. This was one newspaper that I wish I had never seen. It caused me to become injected into a religious controversy that came near meaning my finish.

This newspaper appealed to me, why I do not yet know, but I believe the style was more fascinating than any newspaper I had ever seen. I ordered 400 copies of a special S. O. S. (Save Our Schools) edition. In that edition the Papacy at Rome was pictured as taking over every school in America.

In the dead of night I placed a copy of that paper on every man's porch in the mining town of Cannelton.

Talk about seething! The town boiled for a few days. In some way the Catholics and Knights of Columbus found out that I had distributed the papers. My scalp was to be taken.

Friends came to me and said: "Man you will get killed."

"Why," I inquired, "don't I have a right to distribute literature if I wish?"

"Yes, but you are stirring up a religious controversy and it is dangerous," they warned.

On the following Sunday morning I learned that my friends knew what they were talking about.

Three Italians came into my house while I was sitting at the breakfast table. One of them did all the talking. His talk consisted of abuse. I ordered them out and they refused to go. I proceeded to throw or knock one of them out and while I was doing that another of them leaped upon me with a knife. I secured a firm grip on the hand in which he held the knife and he was not able to cut me vitally, but I was compelled to put him away or he would have worn me down and his friend was getting ready to come to his rescue.

This was the outcome of the only religious controversy in which I have ever become involved. And never again. Religious controversies concerning what form of worship one shall adhere to are silly, asinine, and contemptible. The man who will stir up religious strife and encourage intolerance is more dangerous to society than is the anarchist, the bomb thrower, or bolshevist, because history proves to us that rivers, seas, and oceans of blood have been shed over religious controversies. Kingdoms have been rent in twain, empires caused to fly at one another's throats, turn animal and rend like beasts of the forest and birds of prey, and all that they have accomplished is a bloody record to which the atheist, agnostic, and unbeliever can refer.

The pity of it is that there can yet be found a few men with hummingbird mentalities that will occasionally strike a match against this powder magazine, because if there is one paramount issue over which human beings can be incited to slay and kill one another that issue is religious controversy. None of it for me. A man can worship his maker under his own vine and fig tree and according to the dictates of his own conscience. As for me, I want to do as I please, ever mindful of the fact that my religious, civil, and political liberty ends at the point where the liberty of the next human being begins, giving each of them all that I take for myself.

New Year's Day 1916 found me facing a multitude of discrepancies. I had worked but a few days in six months. I was heavily in debt. I was cited for trial in January for my affair with the Italians. Lawyers had to be paid, etc. But I went to work and never lost a day until trial day.

The trial was of short duration. I used only the state witnesses. The jury deliberated only 26 minutes. The verdict was "Not Guilty." I was free again but $452.00 in debt. By June 30, 1916, my debts were all cleared up and I had saved a little money. My wife and I spent the Fourth of July visiting my parents.

## Chapter V

My first wife died on September 16, 1916, leaving me with three boys, the oldest six years and the youngest only six months old. My sister came to stay with me until I could regain my equilibrium and decide just what course to follow. We thought my baby was going to die in spite of all we could do. One day she was sitting before the fire trying to pour a dose of doctor medicine into the little fellow's stomach when she said, "I am not going to force another dose of that darned medicine down him if he does die," whereupon she threw the medicine into the fire. The kid improved from that time on.

Letters were pouring in demanding of me that I run for an office in the United Mine Workers organization.

"What will I do?" I asked of my sister.

"Go to it," she advised. "I will take the kids home and you can make that your home again."

I proceeded to take her advice. My service as a mine committeeman, handling grievances which arose between employees and employer, was useful after I accepted the invitation to become a candidate for an office in the miners' union.

I visited Eskdale, the residing place of Frank Keeney, Lawrence Dwyer, and others, who were vitally interested in building an organization in West Virginia. We conferred as to who should accept the nomination for this office and that. Keeney and I were being nominated for various positions and it seemed as if either of us could take political chances for almost any honor the miners of District 17 were able to confer.[1]

Obe Clendenin suggested that I run for president and Keeney for secretary, but at this point Dwyer offered a suggestion. He said, "I believe Keeney is the stronger man and if elected would be better fitted to fill the office of president because of his experience in the secession movement."

To this arrangement I agreed. Keeney was to accept the nomination for president, I for secretary-treasurer. So it was agreed and Dwyer further stated, "If we can get you and Keeney elected as president and secretary-treasurer, we don't care who is elected vice-president and executive board members. You and Keeney can hold the others in line."

The campaign was on. Keeney and I were slandered by every method known to political tricksters. The old ousted officers tried in vain to excuse their previous malfeasances and double dealing by attacking us. But Keeney and I knew how to hit back, and every time we replied to their innuendo we left them prostrate in the dust.

They pointed out that Frank Keeney bolted the miners' union and led a secession movement in an effort to drive the miners' union out of West Virginia. Keeney hit back saying, "Yes, I led a secession movement against the most degraded group of crooks, drunks, and double dealers that was ever known to infest the body politic of a labor union, but never at any time or place did I or my associates do otherwise than swear our allegiance to the miners' union."

The ousted officers attacked me by saying that Fred Mooney lured a poor Italian into his home and shot him in the dead of night. In my reply, I pointed out that the court records were public property, open at all times for inspection, and that if any other group of Italians or ousted officials or any one else believed that I would not follow the same policy as that pursued in the previous encounter, for them to invade my home and undertake to do me injury and learn for themselves.

There were many Italians in the miners' union but this infamous lie had little or no effect. When the mudslinging had reached a successful conclusion and the votes were cast and counted, Keeney and I were declared elected.

Keeney's majority was 343 votes. I was elected by the small majority of 45 votes. Keeney and I had friends to sit in when the votes were counted, and we were notified to report for duty January 1, 1917. Keeney and I were at union headquarters in Charleston when the result of the election was announced. We agreed to have a conference at Eskdale before being inducted into office on January 1.

Elected to the several positions in the organization and to be associated with Keeney and me were a vice-president, five executive board members, and three auditors.

The vice-president elect was William Petry of Boomer. By reputation, Petry was an habitual drunkard, fat, irresponsible, and prone to make "grandstand plays and then pass the buck," as it was termed when one of the officers would assume an untenable position toward controversial questions arising between employer and employee.

The personnel of the executive board was not very encouraging. C. H. Workman, of Kayford, was born and reared in the backwoods of Raleigh County. It was said of him that he once taught school, but to see him in action one was compelled to wonder how he ever managed to teach anyone anything other than how to stand still. He, like Petry, was a "buck passer"; he was stockily built and was not a thinker, for fat men are seldom thinkers. But Workman was agreeable and after securing the instructions of his associate officers would proceed accordingly.

Ed Scott, "Fighting Ed," he was termed, was tall, raw-boned, and was or had been an athlete. In appearance he resembled Frank James minus the beard Frank was wont to wear. Scott was master of a hundred brawls; he could fight. Of ability, he had none, and when under the influence of liquor he was considered dangerous.

James Diana, an Italian by birth, small in stature, was capable and industrious, speaking both Italian and English fluently. Diana had served as an executive board member under previous administrations.

Nick Aiello, another Italian, was young, energetic, of good appearance; incapable, yet willing to learn.

W. S. Reese (colored), like Diana, had served as an executive board member under the old regime. Reese was conscientious, capable, and reliable.

Some of those elected held malice toward Keeney but were friendly toward me. Others were tolerant insofar as Keeney was concerned but entertained bitter feelings toward me because of the word battle which had ensued prior to our election.

On January 1, 1917, Keeney, Petry, and I assumed the duties of office. I notified the board members to report on the 5th, I believe. Keeney and I discussed the possibilities of organizing the executive board and starting it to work. I said to Keeney, "Well, we are here, now what are we going to do about it?"

"We will get them together and see where we stand," he answered. Rumors were afloat that some of them were antagonistic to each of us and that their expressed intentions were to the effect that they were going to throw us out through the window. "What do you know about Ed Scott?" Keeney inquired of me.

"Scott is not afraid of the devil himself," I said, "and will go anywhere at any time, but he is a drunkard and I fear he will cause trouble."

"Well," Keeney drawled, "those that refuse to comply with the working rules and regulations will have to get out and we will appoint men in their places who will outline policies and work in harmony." So we agreed that no matter what happened, the two of us would stick together and that we would organize that body of men and start it functioning even if we were compelled to use strong-arm methods to do so.

On the morning of January 5, 1917, Keeney kissed his wife and babies good-bye and I did the same with my boys. We fully expected to have some serious trouble before the end of that day. However, when the men arrived and greetings were exchanged they saw that Keeney and I were together and they decided to play the game. Keeney delivered his message to them, outlined the work expected of us and the dangers with which that work was fraught. "Gentlemen, I wish to know now if you are here to do the things required of you or do any or all of you feel that you have a personal axe to grind?" he inquired of them. "If you have an axe to grind, let us hear from you now. On the other hand, if you are willing to go along with us and build an organization of which we will all be proud, then let us know it and we will get down to business."

Confidences were exchanged and ere long a friendly feeling existed. At that time there were about 8,000 members in District No. 17 and $7,000.00 in the treasury. Those newly elected officers organized and started to work. It was decided to fill in where loose links appeared within a certain circle and complete that circle before any effort was made to expand. This we proceeded to do until new and untouched fields only remained.

During this campaign, Ed Scott and Nick Aiello were dispatched to Red Warrior² on Cabin Creek to organize a local union and install officers. When they arrived at their destination, Scott was drunk and he proceeded to shoot up the town. Aiello reported to the office the next morning and said, "Boys, I won't go out with that man again for he is going to get into serious trouble and drag someone else in with him."

"Why not prefer charges against him and impeach him from that office?" I asked.

"Me prefer charges against that man," he blurted, "when he vows he will kill the man who does it?"

Scott had been friendly with the ousted officials and we knew it. So we checked him while in the city and learned that he was conferring with and being tutored by them. I said to Keeney, "We must act or this fellow is going to disrupt the entire procedure and bring us into disrepute."

"You know him better than I," Keeney answered, "You know what it may mean."

All that day Scott had been pacing to and fro, from one office to another, drunk, nervous, and fingering his gun. At times he would walk quietly up behind me while I was at work at my desk. His presence eventually became unbearable so I turned around to the typewriter and wrote out charges in duplicate against him. I signed both copies, handed one copy to Keeney and walked into the assembly room where Scott was sitting and spread the other before his eyes. He remained seated while reading the charges, then he arose and left the building. He did not appear for trial, so the case was decided against him and his position declared vacant.

Some time after Scott left the office, he called me on the phone and said, "I am going to meet you on the street and shoot it out with you when you come out."

"I'll be down right away," I replied, reasoning that if this ordeal had to be gone through, it might as well happen then as

later. But after walking around several blocks and seeing nothing of my adversary, I returned to the office and resumed work.
W. F. Ray, of Black Betsy, was appointed to fill the unexpired term of Scott. Ray was of typical native stock, fiery and totally blind to everything but right. In addition to possessing the qualifications of an organizer and an executive, he was friendly to Keeney and me. He had been active in the convention of 1914, when the old officers were discredited and eventually ousted by the international union and the autonomy assumed by the parent body.

\* \* \* \* \*

Following the appointment of W. F. Ray, we began to organize in earnest, reaching out and getting those miners adjacent to or near the main hub, as we termed the Kanawha Valley. Acting upon the signed request of a group of miners at Bower, Braxton County, and at Gilmer, Gilmer County, we dispatched P. F. Gatens to those mines to make an investigation. He reported that the miners were ready and anxious to organize.

Meetings were arranged and 500 miners were organized into several local unions. Immediately following this action, the miners were discharged and eviction notices were served upon them. Tents were ordered from the international union. Every effort was made to secure a conference with the coal operators affected but to no avail. The strike was prolonged into the summer months with no hope of a settlement in sight. Tent cities were in evidence at Bower and Copen.

Winter was approaching and the strike seemed to be lost so we decided to go to Bower and look the situation over and decide on some definite policy. President Keeney, Vice-President Petry, board members C. H. Workman, James Diana, and I proceeded to Bower. We visited the strikers in their tents at Bower and Copen.

Our investigation at Bower was ended and we were making preparations to return to Gilmer for the night when Russell Shelby, a striker living in the colony at Copen, came to us all out of breath and said, "Boys, if you value your lives, don't start back to Gilmer this evening, for the gunmen are lined up on both sides of the road and are going to fire on you from ambush.

I crawled up near three of them and heard them talking over their plans." "Soon after you left our camp," he added, "we saw them going into the hills with their rifles and reasoned that something was wrong. So I started around the hillside and almost ran into three of them, but luckily they did not discover my presence."

Eighteen or twenty gunmen had been on the premises of these coal companies since the beginning of the strike, with the exception of the Gilmer Fuel Company. The gunmen were under the leadership of one Nathan Ash. Ash had been a gunman for years.

After being informed of the ambush we held a hurried consultation and decided to walk in the opposite direction to Cutlip Siding, rather than return to Gilmer. We had eight miles of railroad ties ahead of us and a tunnel to walk through.

The sun began to disappear below the horizon before more than half the distance had been covered, and the steel rails began to contract as the atmosphere cooled. This contraction caused the rails to crack as if some one were striking them with a hammer. I had sat down to rest when this was noticed by some one in the bunch.

One of the group gave me the wink and said, "Boys, those sounds are caused by an approaching velocipede." It was generally known that the gunmen owned one of these small gasoline operated cars. Board Member C. H. Workman was fat and short-winded, but he suggested that we "run to the first residence." "Come on, boys," he said, "let us reach a place where some one will see us killed, at least."

We ran for several hundred yards, then I stopped and declared, "I am going to rest if they do kill me."

"Come on, man," begged Workman, "do you want to die here where no one will see you kick off?" Then everyone laughed and Workman became angry; he sulked all the way to Cutlip Siding after we told him it was the contraction of the rails which caused them to crack.

People had retired when we reached Cutlip Siding but we obtained food and lodging from one of the residents. We returned to Gilmer the next morning via the railroad. Freight cars were secured and the strikers were moved into Paint Creek.

* * * * *

Lumberport was a little town around which several mines were located. A meeting was billed at this point and Mother Jones and I were dispatched to organize the miners. I was introduced as the first speaker and after speaking for about 40 minutes, outlining the policies and principles of the United Mine Workers of America, without getting an audible response from the meeting, I decided to quit. I did so by introducing Mother Jones. "What is the matter with this bunch, Mother?" I inquired, "I can't move them."

"Yuh damn fool, don't you know that not one of them understands a word of English?" she said, and laughed heartily at my expense. Mother Jones talked shop to them by telling them that the bosses were through kicking them around, etc. We organized five local unions and installed the officers that evening.

When we returned to headquarters at Fairmont, we were worn out and hungry. The Watson Restaurant was a very good place to dine. We entered the restaurant and found a vacant table. "What are you going to eat, Mooney?" Mother asked.

"Don't know yet, Mother," I said, "but I believe I will have some of those French fried potatoes, for one thing."

"Potatoes, hell," she exploded, "order you one of those T-Bone steaks, you have earned it."

I, like many other officers, was skeptical about ordering food when Mother was around for she occasionally gave the officers "hell" as she termed it, "for filling their stomachs with beefsteak at the expense of the wretches." Each of us put forth every effort to avoid Mother's ire. For several years we were successful, but there came a time when we disagreed.

* * * * *

During the summer we had negotiated a wage agreement with the coal operators of the Flemington field in Taylor and Preston counties. This conference was held at the Waldo Hotel in Clarksburg. Only a small group of coal operators was affected and they entered into that wage agreement under pressure.

From 30 to 40 gunmen were in the lobby of the hotel all day each day that conference was in session. Every few minutes someone would call for one of the coal operators with whom we were dealing, and when he would answer the phone it would be one of the gunmen. "Why don't you come out of there, you coward, and fight the union like the rest are doing?" the gunman would say.

The operator answering the phone would only grin and resume work towards consummation of a wage contract. "Those gunmen see their jobs slipping," the operators would say as each in turn took his medicine over the wire.

Those gunmen of northern West Virginia were organized and hired out by the Shuttleworth brothers; these two brothers were thought to be the king bees in many counties. Andrew Watkins, international board member of the miners' union from Ohio, had felt the fangs of this gang. He was blackjacked while crossing the bridge that leads to the depot in Clarksburg and left for dead.

The gunman who struck Watkins used a coupler hose from the end of a railroad car. Watkins' head came in contact with the iron on the end of the hose. His skull was fractured and for several weeks he lingered between life and death but eventually recovered. Ira Marks, a member of the executive board of District No. 17, United Mine Workers, was also blackjacked while in the city of Clarksburg and almost killed. It was several weeks before he recovered from the rough treatment accorded him by the gunmen.

John J. Cornwell was then governor of West Virginia. The officers of the miners' union called upon him to assist in the elimination of the gunmen. Early in 1918, he agreed to do everything in his power to induce the coal corporations to stop the activities of the gunmen in their employ. Cornwell was a Democrat and he was indebted to the union men in Kanawha County for his election. There were then about 3,500 votes in that county that were hard to control. Cornwell had been elected by a very small majority, and that majority was cast in Kanawha County.

When he saw how close the election was going to be he came to these men and said, "Boys, throw your support to me and I will do everything in my power to remove the gunmen from the state." In a measure, John J. Cornwell carried out his promise. However, the bludgeon system was so deeply rooted in the political, civil, and industrial life of the state that only a superman, such as Oliver Cromwell or Napoleon Bonaparte, would have been able to prevent its functioning.

"Be it enacted" would not suffice; words had been tried and found wanting.

Governor Cornwell informed us that Clarence Watson wished to be elected to the United States Senate from West Virginia. Mr. Watson was the controlling factor in the Consolidation Coal Company. We were advised to interview Mr. Watson at his residence in Fairmont.

Along with Frank J. Hayes,[3] then vice-president of the international union, we called upon Mr. Watson. We suggested to him that withdrawing his opposition to the miners' union and permitting them to organize during the political campaign might be good political expediency and cause thousands of votes to come his way. At that time we really believed that this would be the case. But lo and behold, it had just the opposite effect.

Imagine 23,000 coal miners under the "iron heel" of the gunmen system for 20 years; think of the crushed hopes, smothered ambitions, cracked heads and the thousands of insults to which they had been subjected at the hands of the bludgeonists. Then say to them, "the gunmen are going to cease cracking heads; you can now have the freedom guaranteed you by the American Constitution." Permit them for the first time in their lives to enjoy the civil liberties of which they had been deprived for two or more decades, and expect them to give you their votes in return. It was too much to expect; they just didn't do it.

By the time the election rolled around we had organized 22,000 men and secured a closed shop working agreement for them.[4] Among them were men from many countries. Faces from the Steppes of Russia, from Rumania, Italy, Turkey, Greece, Poland, Armenia, and many others were included. I assisted in organizing one local union at Monongah in which about 27 different languages were spoken. They had come to America as immigrants looking for a land in which a higher degree of freedom prevailed than in their own country. What did they find?

In the countries from which they came (many of them lured by the lies of foreign employment agents), there was no such thing as political action. Their only known remedy when oppressed was to throw a bomb, a knife, or to fire a bullet.

They were turned loose on the body politic of America with no training about our form of government and the operation of our institutions. They were ushered into the presence of a system that gave forth an aroma of the feudalistic days. Instead of being encouraged and educated in the functions of our government, they were browbeaten and kept in ignorance and submission. Penalties have been and will be paid for this neglect. During the transitional period in the coal fields now under discussion, it was pitiful to see the demonstrations of ignorance as far as the ideas of trade unionism were concerned.

It became the duty of the miners' union to educate these men as best it could. The change from industrial tyranny to that of freedom obtained by a joint agreement under the competitive bargaining system was new to them and they did not fully comprehend it. Branch headquarters were established in Fairmont and a force of competent men was kept in readiness at all times.

Early in that campaign of organization, we came in contact with some pitiful objects in the form of cripples, men who had contracted occupational diseases, etc. Ira Marks and I were sent to Berryburg to meet with the miners employed there. When the meeting convened the superintendent and the mine foremen were present. Time arrived to administer the obligation and the entire management wished to join the union along with the miners. It became my painful duty to inform them that they were not eligible for membership. But a splendid spirit of fellowship prevailed, despite the fact that the officials of the company were not permitted to join the union. An Austrian came to me after the meeting. His entire body was drawn into unnatural shapes by inflammatory rheumatism. He said to me, "Looka atta me, for 20 years I worka ina water upa to my knee, some time deeper, now all I gotta is wife and sixa children. They no baila the water from my place, now I no gooda for nothing. Buta thisa be gooda for everybody else but no helpa me now."

Others were legless, armless, some drawn double from broken backs, but despite their infirmities they were for the union, first, last, and all the time. The attitude of the coal miners in that field when the pressure was withdrawn proved conclusively to me that there is no group of workers in any industry that cannot be organized into a trade union if the visible and invisible influences of employers are withdrawn or even minimized. The truly voluntary educational process of trade unionism has wielded a powerful influence in stabilizing many of the major industries.

By late fall of 1918, more than 20,000 miners were organized and enjoying the full benefits of a closed shop working agreement. That agreement included the essence of more than a dozen agreements, and during the period of its operation few misunderstandings were encountered relative to its interpretation. The worst enemy of joint wage agreement in any community or industry is illiteracy. Ignorance, teetotal, dumbfounding ignorance is the supremely remarkable thing with which a labor leader comes in contact—the inability of his constituents to understand joint relationships—their inclination to look upon and regard a working agreement as if it were made only for their benefit.

## Chapter VI

Early in 1919, after harmonious relations were established in the twelve and one-half counties in northern West Virginia, we began to look toward Logan, Mingo, McDowell, Wyoming, and Pocahontas counties in southern West Virginia.[1] John J. Cornwell was yet in the gubernatorial chair. He had rendered valuable assistance in working out an amicable situation in northern West Virginia. Therefore, we approached him relative to the southern counties. In that conference were President C. F. Keeney, Frank W. Snyder, editor of *The West Virginia Federationist*, James Riley, then president of the West Virginia State Federation of Labor, and myself.

"What will you do toward helping us work out an arrangement in Logan, Mingo, and adjoining counties?" we inquired. He tried to dissuade us from any effort to organize the miners in that territory. But these fields were constantly being referred to by the union operators as their direct competitors, and we assured him that we were going to undertake to organize them at all hazards. So it was agreed that he be given 30 days to make an effort to have the coal operators withdraw their opposition and permit the miners to organize.

We separated that evening all in good spirits, and immediately following that conference the executive board of District No. 17 was called into session. For some reason, real or imaginary, President Keeney asked the board to permit him to commission 50 organizers to enter Logan County. I did not then and do not yet know why the change in policy occurred almost overnight. I asked the question before the executive board, "Are we going to take this action in the face of our understanding with the governor?"

I was informed that "this organization has reached such proportions that it is no longer compelled to await the whims of every politician in the state."

Keeney was mad for some reason. Prior to this meeting with the governor, we had produced nine victims of the gunmen to him. Some of them were from Logan County, some from Mingo and McDowell counties. They were beaten and maimed and some of them had been left for dead. One miner had been shot in the head, but his wound was not fatal.

Don Chafin was then sheriff of Logan County. The coal operators of Logan County did not employ detective agencies to keep out organized labor. The sheriff appointed the men employed by the coal companies as deputy sheriffs and they operated as officers of the county and state.[2]

The organizers were selected and commissioned as President Keeney requested. They boarded the train at Charleston and went into Logan County via Barboursville. When they arrived in Logan a special train met them. On this train were the picked gunmen of Logan County. They were equipped with machine guns, rifles, and revolvers. The organizers were given the privilege of returning to Charleston on the next train. Outnumbered five or six to one, they took the advice of the gunmen and amid much abuse and innuendo they departed for headquarters at Charleston. Each of them returned his commission and sought his home.[3]

For months prior to the time these men were commissioned as organizers, there had been considerable criticism throughout the district because the officers did not organize Logan County. When the time came to select the organizers, Keeney asked me to assist him in the choice. We selected every critic whose name we could remember. "Give them a try at it," we said, "let them find out for themselves just what the conditions are and what the opposition consists of in Logan."

As a result of this adventure, criticism ran at a low ebb for many moons. Several days after the organizers were given until train time to leave Logan County, I met Governor Cornwell on the street near the statehouse. To say he was angry would be putting it mildly.

"What in the devil do you fellows mean?" he inquired of me. "Why didn't you keep your promise made to me in that conference?" I frankly informed him that I did not know. I knew at the time that he thought I was lying, but I was telling him the truth. He almost stormed at me, brandishing his cane, but I again assured him that I did not know why his confidence had been betrayed. "You fellows have received all the assistance from me that you will ever get," he asserted.

The march of events was convincing proof that he meant every word of that assertion. However, the flames of discontent were being continuously fanned by the acts of violence of the gunmen.

\* \* \* \* \*

During the first days of September, 1919, the minds of the union miners became so inflamed that they began to congregate at Marmet with the avowed intention of marching on Logan County.[4] Prior to and in contemplation of this venture, those who did not already possess guns and ammunition purchased such as they could secure and agreed upon Marmet as the point of bivouac. Many of their leaders were constant visitors at union headquarters with pleas for help such as food and supplies. They were constantly denied these requests by the officers of the organization.

While they were encamped near Marmet, Governor Cornwell signified a desire to address them. They in turn requested him to appear before them. It fell my lot to ask the governor to speak to them. I was not a little apprehensive concerning his going before them at night. However, he insisted upon the venture.

Among that gathering were six or seven hundred ex-soldiers. They had returned from service in the United States Army disillusioned concerning the democracy they had fought for overseas, sweated for in cantonments under semi-tropical and tropical suns. Army equipment was in evidence, such as regulation clothing, helmets, and gas masks. Negroes, Italians, Hungarians, whom the bigoted and hypocritical four-flushers of the nation referred to as "foreigners," were sadly in the minority.[5]

Upon the arrival of Governor Cornwell, who was accompanied by Mrs. Cornwell, his private secretary, Mr. James W. Weir, and Mrs. Weir, he was given an automobile to use as a platform.

He asked them, "Boys, do you not know that every one of you is acting in violation of every law against bearing arms and that your are taking the law into your own hands?"

Many of them replied to him in unison, "There is no law in West Virginia except that decreed by the coal operators, and you know it, governor."

He explained: "The gunman system is a condition that I did not create and a system to which I am opposed, and I assure you that my good office shall be used during the time I remain governor in an effort to eradicate this system from your state." Hundreds of times these same men had listened to this same promise made by politicians of both parties. They knew that these promises were only intended to soothe their state of mind and lull them back into a temporary state of contentment.

When the governor had concluded his address only one burly Negro said to him, "Mr. Governor, you made a nice speech, we likes yo' talk, but it don' mean nothin', and I'se erfraid you done lose." His remark apparently voiced the sentiments of the entire gathering, for there was much applause.

Five or six thousand men listened to the governor's speech. Camp fires, miners' lamps, and auto headlights threw glinting rays upon the steel of rifles, side-arms and shotguns. When the governor had turned his automobile around to depart for Charleston, some one in the crowd proposed a salute to the chief executive. Keeney and I started to protest but our voices were drowned by the roar of more than three thousand shots fired into the atmosphere. Rifle, shotgun, and pistol bellowed their salute of respect, or defiance, who shall say?

"You will frighten those women," we remonstrated.

"Hell of a note if we can't fire a salute to our own governor, when during the war we were forced to salute every jackass that happened along just because he had big ears," they answered.

Immediately after the governor departed, the cry of "On to Logan" was taken up. Camp was broken and the marchers began to move in that direction.[6]

❋ ❋ ❋ ❋

The "Miners' March" of 1919 was the beginning of the disintegration of the district organization. Immediately following this occurrence, the elements of suspicion entered the ranks of the officers and internal clashes concerning policy and procedure became frequent. The district had been divided into four sub-districts: No. 1 with headquarters at Montgomery; No. 2, at St. Albans; No. 3, at Grafton; and No. 4, at Fairmont.

William Blizzard was then president of Sub-district No. 2. Blizzard was all fire and dynamite, hot-headed and at times irresponsible. In addition to the aforementioned faults he was a mimicker. His imitative tendencies were unsurpassed but he was incapable of measuring the potential outcome of his many irresponsible plunges. For some reason, then not known, President Keeney was inclined to take sides with Blizzard regardless of the intangibility of his positions. Blizzard spent most of his time at the main office in Charleston, "watching the officers so they can't put anything over on us," he was wont to inform his secretary.

When the increase awarded the miners by the Bituminous Coal Commission of 1919 was to be applied, I was selected to represent the district organization on a uniform rate commission. Percy Tetlow,[7] of Salem, Ohio, represented the international union and he and I worked together along with the representatives of the coal operators for several months in an effort to eliminate the inequalities in tonnage rates, prices then effective in the various seams of coal, and those existent in one community as against another.

When our work was completed and the results announced, Blizzard "went up in the air" as the miners termed some of his tantrums. In the Coal River section, part of which territory is in Boone and part in Raleigh County, the increase was taken from coal fields where rates were a few cents higher and transferred to the section over which Blizzard presided as an officer. He assumed the position that the increase should have been applied to the high rates and the deficiency in "his" territory made up by an additional increase to be given by the several operators affected. Immediately throughout the affected part of the subdistrict where Blizzard had subordinate jurisdiction, dissension began to appear.

A mass meeting was called at Whitesville and Keeney and I were invited to appear and explain our position. The theatre building had been arranged for and approximately seven hundred men, women, and children were at the meeting. Mother Jones accompanied Keeney and me on this occasion and in many ways the meeting was like most of its kind. However, it was apparent to the casual observer that bitterness was rife among those miners. I had concluded my explanation of the work done by the commission when one Ed Scott, mentioned in a previous chapter, entered the theatre by a side door. Scott walked to the stage and I extended my hand to him and assisted him in mounting the stage.

In the meantime, I had introduced Mother Jones as the next speaker. By intuition or otherwise, I knew that Scott was there for some definite purpose and that purpose seemed to concern me. He kept shifting for a position behind me and I watched him keeping his hand under his coat as though gripping a weapon. However, I was "heeled," as many term toting a gun, and I knew just how to use one. Several people in the audience kept looking at me and nodding in the direction of Scott as though to warn me. But for some reason I had never feared Scott. I had pushed the safety clear on my automatic and was holding it in readiness when there came an interruption which relieved me of any further participation in the controversy.

Before entering the theatre, Scott had been heard to assert that "there are two men in that meeting on whose heads there is a price of $5,000 each and I am going in to collect." Tom Akers, the local constable, was informed of Scott's threat and ran to the office of the justice of the peace and secured a warrant for the arrest of Scott. When Akers entered the front door, he strode straight for the stage. "Give me a hand," he said, and I reached for him. As I pulled him upon the stage, he called to Scott, "Throw them up, Ed, reach high." Scott went for his gun and even before Akers was through calling on Scott to surrender,

Akers' gun cracked and Scott emitted a groan. During this episode Mother Jones kept speaking and only once did she turn around and that was when the shot was fired.

I yet had Akers by the left hand when his gun exploded from his right. Letting loose of his hand I stepped to the edge of the stage and began speaking to the crowd in a loud voice. "Sit still in your seats and do not become excited for everything will be cleared up and explained in a moment," I said.

Not one of the audience undertook to make an exit and after moving Scott into an anteroom and sending for a physician, we went through with the meeting to a successful conclusion. The avowed intention and the attempt of Scott to carry out his threat solidified the people at that meeting and hundreds of them shook hands with Mother Jones, Keeney, and me before we boarded the train for Charleston. "Stay with them, boys," they admonished, "there is something good in you two or they would not be trying to kill you." Several weeks later I was told that William Blizzard had encouraged the dissatisfaction which had grown out of the adjustment made in prices. Their grievances were only imaginary, inasmuch as they had not been properly explained.

* * * * *

The spring of 1920 marked the beginning of events that were eventually to cause the downfall of the miners' union in West Virginia. Many grievances were brought up from Sub-district No. 2, of which Blizzard was president; the majority of them were indefensible and unreasonable insofar as the terms and conditions of the joint working agreement were concerned. As chairman of a joint board meeting consisting of an equal number of representatives of the Kanawha Coal Operators' Association and the miners' union, before which all grievances not settled locally or with the commissioner or his assistant were brought for final adjustment (unless a case was referred to arbitration), I was compelled to adjourn a meeting sine die in order to prevent William Blizzard and B. F. Morris, assistant commissioner for the coal operators, from engaging in a gun battle.

The dispute arose concerning a dead work scale at Jarrolds Valley. On one side of the mine there appeared a band of slate or other unmarketable impurity in the seam of coal. This band was several inches in thickness. A price was agreed upon for separating this band from the coal, said price to remain in effect until such time as the seam of coal returned to its normal condition. The impurity or slate "ran out" or disappeared, and the operator affected asked for an adjustment and relief from paying for something that did not exist. Blizzard denied the operator that adjustment. The operator affected appealed his case to the joint boards of the two organizations and asked for relief. President Keeney asked me to assist Blizzard in the adjustment of this case and several others that bore the same marks of untenability. Blizzard and Morris became engaged in a heated argument and the lie was passed. Morris said to him, "Bill, you can't talk to me like that and get by with it," and left the room. He returned several minutes later and I acted upon the presumption that he had heeled himself with a gun and adjourned the conference. I was later told that my surmise was correct.

B. A. Scott, who was then international board member from District No. 17, was in that conference and is thoroughly acquainted with the details. During the argument which ensued while the grievance was under consideration, Haydon Morton, the affected operator, said to Blizzard, "Bill, I will operate my mines non-union before I will stand for such as that."

"Scab your damn mines and see if I care," Blizzard retorted.

Morton did operate his mines on a non-union basis, beginning soon after that conference.

Blizzard made insulting remarks to me after that conference adjourned. Sufficiently so that I informed him, "Insofar as your grievances are concerned you can secure someone else to assist hereafter."

He was arrogant and irritable, possessed plenty of nerve or gall, but like "Simpson's Bull" or "Collins Ram" he used no judgment whatever. He had no conception of the consequences of an act committed as a means toward an end. He committed the act and let the consequences of his act fall upon whom or whatever they might. He seemed unconcerned.

For several months previous to these occurrences, powerful human forces known and unknown had exerted themselves to create a friction between President Keeney and myself, but to no avail. We had discussed several of these forces, their methods, etc., and through a perfect understanding we had eliminated them from the stage of action. But Blizzard impressed me at times as though he were envious because he thought Keeney and I understood each other and worked together in harmony at all times. Keeney's devotion to Blizzard eventually caused us to disagree on several occasions. In spite of that, I do not condemn, for who shall explain a man's love for his comrade, his loyalty to a friend. Love or hate must write the summary of these things; I do not feel capable of the undertaking. For often in reverie I form a picture in my mind of the empire that could have been built from the material we had on hand when dissolution entered the game. Every man has a weak point in his composition. With one it may be egotism, another liquor, women, gambling, a friend, or something else; but woe unto him when that weak point is discovered by the enemy, if he has one, and if he stands in the way of the ambitions of that enemy.

✻   ✻   ✻   ✻   ✻

The miners of Mingo County became restless early in 1920 and following an announcement of a policy of reducing wages at Sprigg and several other mines on a certain date, the coal companies posted notices to that effect. Several hundred miners immediately went out on strike and sent a committee to union headquarters at Charleston with petitions to the effect that they desired to become members of the miners' union.[8]

"Are the people who signed these petitions willing to return to work until we can get to them," the committee was asked.

"They are if the coal companies will re-employ them," we were informed. The companies refused to re-employ many of the miners and several days later men were dispatched to organize those men. Following the organization of these miners, wholesale evictions began to take place. Pursuant to this action unlawful detainer notices had been served giving the miners their choice of renouncing the union or vacating the property of the affected coal companies at the end of a period of three days. At Stone Mountain near Matewan many families were evicted, their few household belongings piled out beside the public highway. The Baldwin-Felts detectives (in true Stanaford Mountain, Holly Grove and Ludlow style[9]) appeared on the scene.

Immediately following the organization of several groups of miners in Mingo County, C. E. Lively[10] appeared upon the scene. Ensconcing himself in a restaurant at Matewan, he began to attend the meetings and to make speeches along with the miners' officers. He was loud in his denunciation of the gunman system and advised the miners to join the union and fight for their rights. His restaurant was a rendezvous for the strikers. On May 6, 1920, William Blizzard and I visited Matewan and in the evening we spoke to approximately 3,000 spectators, most of whom were strikers. The rain fell in torrents previous to and during that meeting. Someone held an umbrella over each of us while we addressed the gathering. Whisperings were then abroad that the Baldwin-Felts gang was coming into the strike zone in full force. We outlined the policies and principles of the United Mine Workers as we understood them. We admonished them to stay within the confines of the law but to prevent the private gunmen from superseding the duly elected and qualified officers of the law as we had known them to do in other localities. At the conclusion of this meeting and after we had retired to the restaurant operated by C. E. Lively, he became insistent that I spend the night at his home. Lively and I were born and reared in the same community and at one time belonged to the same local of the United Mine Workers of America. However, somewhere in my subconscious mind there lurked a suspicion that all was not right with Lively. Acting upon this "hunch," I did not go to his home.

Sid Hatfield, "Two-gun Sid" the press tabbed him, was a native of Mingo County, a descendant of the pioneers who had settled western Virginia. Sid Hatfield worked in the mines when coal was discovered and mines opened up. He was six feet in height, weighed about 165 pounds, had strong shoulders, high cheek bones, a dark complexion, and a smile that would not come off.

Prior to the strike in Mingo County, the citizens of Matewan had selected Cabbell Testerman for mayor. He was a businessman, owning a jewelry store and soda fountain. When elected mayor, he selected Sid Hatfield as his chief of police. Testerman was of short stature, fair complexion, and weighed about 170 pounds.

When the Baldwin-Felts detectives under the leadership of Albert and Lee Felts arrived in Matewan, they called upon Mayor Testerman.[11] "We wish to make the town our headquarters while down here and to mount some machine guns on top of some of the brick buildings in the town, with your permission of course," Lee Felts announced.

"No objections to your making the town your headquarters just as long as you wish to stay," stated the mayor, "but what is the idea of mounting machine guns on top of the buildings?"

"So we can command the surrounding community," replied Felts.

"You cannot mount machine guns anywhere in this town and as long as you are here, you are going to conduct yourselves in accordance with the law," asserted the mayor.

Albert and Lee Felts were accompanied by C. B. Cunningham, Booher, and nine other operatives, all Felts employees. Chief Sid Hatfield had been a silent listener up to the time the mayor refused the request of the detectives, but he now decided it was time he had something to say or "horned in" as he termed it. "What do you aim to do here, Lee?" he inquired of Felts.

"We are going to evict these miners from the coal company houses," he replied.

"Accordin' to law?" asked the chief.

"Well, some might not think so."

"I have nothing to do with what you do outside the town limits but some of those Stone Mountain miners live inside the corporation and you are not going to throw their household goods out without due process of law, not while I am chief of police."

"Then maybe you won't be chief very long," warned Felts.

"Whether I am chief 15 minutes or 15 years, you and your men will abide by the law while in this town or in the calaboose you will go."

"We'll see, we'll see," purred Felts.

Turning to the mayor, Felts said to him, "We can make it attractive to you if you will do as we say."

"In what way do you mean?" inquired Testerman.

"You can name your price if you will let us operate unmolested while here," he offered.

"I am not for sale gentlemen, and you can consider the interview closed," the mayor said. The detectives filed out growling threats at the mayor and his chief.

"Well, that's that, and where do they go from here," said the chief as he cut an enormous chew from a plug of tobacco and strolled out the front door. The detectives had gone toward the north end of town, and as the chief walked down the sidewalk someone came running from the direction taken by the detectives with the information that they were going to "throw out furniture."

Chief Hatfield strode on down to where they were talking to a miner about dispossessing him and inquired, "Felts, have you got an order from the court giving you authority to evict these miners?"

"No," Felts replied, "I do not need one."

"You will need an order before you evict anyone and you had better let them alone until you secure such a writ."

Felts became infuriated at this and told Hatfield, "You better mind your own business, fellow." Felts issued this warning with quivering lips, and his eyes drew into mere slits.

"I'm minding my own business and I'm telling you for the last time not to try it." With this Hatfield turned aside and returned to the bank where many citizens had congregated to see the outcome of the controversy. "Folks, I want 40 or 50 of you to act as deputies. Get your guns and I will swear you in. For those fellows are going to start something directly and we never can tell from what angle it will come."

Ed Chambers was the only policeman in Matewan. He had been appointed by Chief Hatfield to act when his services were required. He was the young son of "Daddy" Reese Chambers. "Daddy" Reese was from North Carolina. Ed was a good-natured, easy going lad; just the opposite of his chief, he appeared harmless and never had much to say. He believed in action.

Soon after a group of citizens had been deputized they and Chief Hatfield were standing in front of the hardware store adjoining the bank. Felts and his associates came up to him and broke the following news, "We are going to place you under arrest."

"For what?" inquired the chief.

"For interfering with the law," Lee Felts informed.

"That's rich," said the chief as his face broke into a broad grin.

Some bystander ran to the mayor's office. "They are putting Sid under arrest," he panted.

The mayor did not stop for coat or hat but walked down to where the detectives, the chief, and 40 or 50 citizens were congregated. "What is going on here?" he inquired.

"We are going to arrest Chief Hatfield for interfering with the law," Lee Felts answered.

"The chief has not interfered with any law and you cannot arrest him and take him out of this town. We will give bail for him, if there is a charge, but you cannot arrest him and take him away," said the mayor.

Lee Felts was standing near the mayor when he delivered his ultimatum. "Oh, can't we?" gritted Felts, and he leaned forward on tiptoes and shot from his side, the bullet striking the mayor in the abdomen. The rise to tiptoes must have been a signal agreed upon between the gunmen for immediately they went into action. In less time than it requires to write two lines of this narrative, ten men were dead or dying and two were wounded.

Chief Sid Hatfield was watching every move made by the gunmen and the bullet fired at the mayor had not more than come in contact with the mayor's body when a bullet from a .44 Smith & Wesson fired from the left hand of the chief struck Lee Felts between the eyes. Almost simultaneously a

.45 Service Colt appeared in the right hand of the chief and a .45 slug plowed its way through the brain of Albert Felts. Those who had been deputized were in action and men fell as a leaden hail found lodgment in human bodies. No word was spoken by either side for both factions had their orders: "Shoot to kill."[12]

Mayor Testerman was rushed to the hospital at Welch and died there. The bodies of the detectives were searched and upon them were found instructions to proceed to Matewan and "get" the mayor and chief of police, no matter what it required.[13]

I was in my office at Charleston when this occurred.[14] C. H. Workman, district executive board member, was stationed at Matewan and he called me over long distance telephone and related the fight. For a few minutes I did not believe Workman but really thought he was kidding. However, he told me to call the Associated Press man and verify his statement. When he said that I knew he was not joking. Lively was sitting in the assembly room when I walked in and broke the news. Several men were sitting around and almost all seemed pleased. The Associated Press representative asked me for an interview, which I sat down to the the typewriter and prepared.

I later interviewed many of the living participants and secured the details from them. I asked several of them, including Chief Hatfield, why so many of these detectives were shot in the head. "They wore 'coats o'nails,' " he replied, meaning coats of mail, "and we were afraid our bullets would not penetrate them."

Indictments were returned against 19 persons who were charged with participation in what was alluded to as "the Matewan massacre," including Chief Hatfield and Ed Chambers, but none was convicted.

The date for trial having been set, the prosecution and defense began their line-up for a real legal battle.

Anderson Hatfield was owner and proprietor of the only hotel in Matewan. This hotel was used as headquarters by the Baldwin-Felts operatives when they began operations in Mingo County.

He was believed to be on the payroll of the Felts agency and was the principal witness before the grand jury at Williamson when the indictments were returned against the Matewan defendants. He was slated to be the principal witness in the trials which occurred later.

One of his chief hobbies was to sit upon the hotel porch and snooze during the evening twilight. The hotel fronted east and toward the Norfolk and Western depot which stood about 200 feet away.

One evening preceding the trials, he was sitting with chair tilted against the wall, hat pulled down over eyes enjoying his customary nap, when a shot was fired from behind the south end of the freight house which joined the depot. Dusk was creeping down the mountainside, and despite the dim and uncertain light the shot went true. The bullet entered his mouth and made its exit through the back of his head.

Sid Hatfield was accused by the prosecution of being the man who shot the hotel owner, but it was proven that the chief made his appearance from another part of town immediately following the shooting.[15]

* * * * *

Among the several tent colonies in Mingo County, Lick Creek occupied first place. It housed approximately one hundred families. Armistice Day, 1920, was an eventful day for Lick Creek. Peace and quiet reigned as the strikers stirred and exchanged greetings at the beginning of another day. But days are never ended for strikers until sunset. The law and order crowd was thirsting for human gore. They must have some excitement on Armistice Day.

About 9:00 a. m., an automobile containing some of the law and order crowd passed through the tent colony, and just beyond the limits of the colony someone fired a shot from the hills and struck the car.

An immediate checkup showed not one striker missing from the colony.

No one was struck by the bullet, but an excuse to destroy the tent colony had been created. The car which had been fired upon never stopped until it reached Williamson, and then its occupants pretended to spread an alarm which had been prearranged. Armed to the teeth, they hurried to Lick Creek.

Alex Breedlove was one of the first men to join the union and when his few earthly belongings were thrown out of a coal company house, he moved into a tent at Lick Creek. Breedlove was coming from his tent when the pseudo law and order crowd ordered him to halt. "Raise your hands!" they cried; "higher!" they commanded. Being unarmed and helpless he complied, and when his hands were as high as he could reach, they shot him dead.

Tents were torn down; pianos were rolled down the hill and demolished. Mattresses, Gramophones, and other furniture and belongings followed.

Liberal saturations of kerosene oil were poured on the wreckage, and matches were touched at several points. As the blaze mounted higher, the pseudo upholders of law and order proceeded to indulge in a war dance similar to the legendary dance of the Apaches when a captive was being burned at the stake.

Mothers sobbed, children whined pitifully while fathers, begging to be spared, lifted their eyes to Him who is always entreated by those who are helpless.

From this scene of desolation, the law and order crowd turned and after throwing the body of Alex Breedlove into a truck, they started for Williamson.

Upon their arrival in Williamson they paraded through the principal streets displaying their "kill." Several of the crowd sat on the last remains of their victim holding his body in such position that his feet dragged the pavement.

The celebration of Armistice Day was concluded with a drunken orgy and each of the participants retired late at night proud and happy that he had made such a wonderful showing of loyalty to his country on Armistice Day.[16]

## Chapter VII

In December, 1920, I asked the executive board of District No. 17 for a vacation.

I wanted to attend the meeting of the Pan-American Labor Congress in January, 1921. I left Charleston on January 4, 1921, along with Mother Jones.[1] We were compelled to stay over a full day in Laredo, Texas, then left Nuevo Laredo, Mexico, about 9:00 p. m. Thursday, January 7. Mother and I were assigned a section on a Pullman car that had been taken from a British railroad concern during the Mexican Revolution several years before. We were shown every courtesy and our trip was uneventful with the exception of glimpses of the scenery as we passed through the cities of Monterey, Saltillo, San Luis Potosí, and Queretaro.

Within about 40 miles of Mexico City, the train stopped quite suddenly. When I looked out there was a string of taxi cabs blocking the railroad tracks. About 40 strikers from a jewelry factory had motored out to meet "Madre Yones." They used red flags to stop the train, then boarded it. Mother Jones was in one seat and I occupied the next seat forward. They threw crimson carnations and blue violets around Mother until only her head and shoulders could be seen. While throwing the flowers they continuously yelled "welcome to Mexico, Madre Yones." When they learned I was accompanying Mother they stuck carnations in my hat band and the lapel of my coat.

We were taken from the train, loaded into one of the cabs, and taken into the city with much yelling and tooting of auto horns. They deposited us at the entrance of the St. Francis Hotel and saw to it that we were safely escorted to our rooms. When leaving, they told us to get ready for attendance at the bullfight that afternoon for tickets would be forthcoming.

Mother told them she did not care to see the bullfight. "But this cannibal will want to go," she added, pointing to me.

We had not paid out a cent since leaving Nuevo Laredo, and when we arrived at the hotel we were given a book of tickets or slips that were good anywhere for anything we wished to purchase in the way of lodging, food, taxi fare, drinks, or any other necessity or comfort. All we had to do when making a purchase was sign and present one of these slips of paper. Mother rested while I attended the bullfight.

The next morning (Monday) about 9:00 a. m., two servants in livery reported to Mother at the hotel and she was told that these two attendants and a taxi cab would be parked in front of the hotel each morning for her use as long as she stayed in Mexico City. They were to take her anywhere at anytime that she wanted to go.

During the day we were taken to a reception where Mother renewed her acquaintance with President Obregón,[2] Secretary Villareal, and others. During the Mexican Revolution, Obregón, Villareal, and perhaps one or two others escaped across the border and were imprisoned in Tombstone, Arizona. There was an attempt made by the then almost defunct Mexican government to have these men extradited. Hearing of this proceeding, Mother Jones hurried to Washington, secured an audience with President Wilson, and prevented their extradition. Had they been extradited under the conditions then prevalent in Mexico, they would undoubtedly have faced a firing squad. So it became evident soon after our arrival in Mexico that nothing in that country was too good for Mother, should she choose to claim it. We were royally entertained every hour of our stay.

During the day we met Morones, Salcedo, Haberman, and scores of others whose names I cannot recall. Samuel Gompers, president of the American Federation of Labor, was also president of the Pan-American Labor Congress. In attendance at the Congress along with President Gompers and representing the A. F. of L. were John P. Frey, Edward F. McGrady, Chester M. Wright, and John Greenway. William H. Johnson, president of the I. A. of M., was also present but I do not remember whether he was with the A. F. of L. delegates or on vacation the same as I. About all the Pan-American countries were represented except two or three.

My greatest difficulty during my visit in Mexico City was in securing the things I wanted to eat. For several days I would go into the kitchen along with the waiters or waitresses and show the cook what I desired. Eventually I invested in a Spanish-English dictionary and learned the Spanish terms for ham, eggs, bacon, oatmeal, coffee, tea, etc. After I had been corrected scores of times on my pronunciation, the food problem was fairly well solved. These trips of mine into the various kitchens, pointing out the articles of food desired and designations of how I wished them prepared, were the occasion for many picnics among the cooks and waiters. However, not being daunted, I laughed with them in each event and by so doing I was accorded much assistance.

Mother Jones and I made use of the cab placed at her disposal on several occasions. We made trips 40 and 50 miles from the city on several afternoons. On these outings we stopped at points of interest. Once we parked at a shop where a Mexican was engaged in carving and scroll work. He was making pieces of furniture that were a pleasing sight to the critical observer. Magnificent was the only word which would properly describe his handiwork. "Look at that," said Mother, "and yet, Mooney, you hear people say these people are no good."

Before we left one little town, we observed an incident which caused Mother to laugh until tears came into her eyes. A Mexican woman had gone to one of the small cantinas after her spouse who had imbibed too freely of mescale, tequila or some other of the intoxicants which were vended in the cantina. She was trying to aid him in reaching home. He was as limber as a wet rag and every time she managed to get him on his feet he would wilt down again taking her along with him. Each time they went down they both sat long enough to have a good laugh. Mother said laughingly, "Will you look at that damn fellow? Every time that woman pulls him up on one side he spills down on the other." They were treating the affair as a joke and finally disappeared into one of the doors, still laughing.

For some reason or other the Mexican workers had the idea that I was Tom Mooney.[3] This was true of the munitions workers and many groups affiliated with the C. R. O. M.[4] Everywhere I went I was greeted as only the Mexicans would have greeted Tom had he appeared in their midst.

One day I went through the National Cathedral and the chamber of horrors. I refer to it as the chamber of horrors because it contained, I believe, every instrument of torture used during the inquisition. There was the thumb-screw, rack, wheel, drip, and the iron coat. It was here that I saw something that in part resembled the monster who had come for me in a dream when I was only 12 years of age. The front appearance of the iron coat reminded me of the coat worn by the devil on the occasion of that dream. However, the instrument was minus the legs, feet, and head of the dream satan. I had read of these instruments in many books, including the works of Eugène Sue,[5] but I had never before seen them.

While standing among these silent evidences of man's cruelty to and mutilation of his kind, my mind wandered back to the many branches of study into which I had delved. Physical, metaphysical, three histories of the world, Spencer, Darwin, Huxley, Havelock Ellis (even including his work on degeneracy), the *Meditations* of Descartes, Redbeard,[6] Volney, and hundreds of others had gone through my cranium. But here, this was not the written word; this was damning evidence against ancient man. He either forgot to destroy this evidence, or he was caught unaware by a catastrophe which struck with such sudden fury that he was not allotted time to obscure his implements of torture and mutilation.

Perhaps, I reflected, it was not intended that ancient man would be permitted to destroy these evidences of his cruelty to his kind. Perhaps fate decreed that they should be left as symbols of his barbarism. Or perhaps they were left to warn future generations concerning what would befall them if they permitted themselves to become lax in their vigilance while the proponents of these ancient superstitions crept upon them and again attained a zenith of power.

A hidden voice seemed to be saying to me, "Forget the chronicles of man; study man himself. Watch his antics in his social sphere; observe the effect of his acts committed as a means toward an end. Analyze his virtues, his faults, and frailties. Watch him in church and again when he goes on the job or enters his business office on Monday morning. Listen to him when he is expounding his virtues as a reformer and then study his acts after he has bumfoozled the 'dear peepul' into trusting him as a political representative. See him dodge, hedge, twist, and squirm to get away from any promises he may have made during his campaign."

My reverie was disturbed by someone patting me on the shoulder and reminding me that the time for closing the institution for the day had arrived. Leaving the museum, I returned to the hotel. On my way I stopped at the Hotel Regis, where Samuel Gompers was staying. He was sitting on the porch resting. "Where have you been now, boy?" he asked.

"Through the museum looking at the instruments used by ancient man to torture his kind," I replied.

"Take a good look at them, boy," he admonished, "and as you go through this life think of the instruments employed by ancient man to force the conscientious objector to believe his way. Never try to force human beings to conform to your philosophies. If you cannot by logical argument and persuasion win people to your side, it is better to leave them alone. Boy, I am a firm believer in voluntarism. If you compel a human being to travel a route he does not understand or force him to do something against his will, when he is no longer compelled to conform to your regulations, he will turn against you. However, once you convince a man that you are right and get him to believe in the principles you advocate without antagonizing him, he will never desert your cause."

This was the only lecture Samuel Gompers ever gave me, but his logic impressed me profoundly. My mind harked back to the hundreds of miners who had been dismembered from the United Mine Workers of America for periods of from six months to 99 years and assessed for fines which upon appeal were found to be unwarranted and unreasonable. This practice was made possible by two clauses in the joint agreement, one of which provided for a closed shop and the other for the collection of fines levied by a local union against one of its members.

In the administration of the affairs of District No. 17, this practice eventually reached the stage where a committee of the executive board was kept in session almost constantly for months for no other purpose than to hear the appeals of members who had been thrown out of the union. In 99 cases out of each hundred, the real cause was found to be personal dislike or prejudice. In every instance where a violation of the laws and principles of the union did not obtain, the decision of the local union was reversed, the member re-instated, and the assessed fine refunded.

Following the lecture by President Gompers, I returned to the St. Francis Hotel where Mother Jones and I were quartered and found her sitting in the lobby. After I had revealed to her my whereabouts during the preceding several hours, we conversed concerning the American labor movement and made comparisons with the Mexican labor set-up. Prior to our departure for Mexico, some discussion had been held relative to organizing in a territory that had not yet been invaded. During this discussion the name of a man had been mentioned as a possible district organizer. Mother mentioned this incident during our conversation in the lobby that evening. "Mooney," she cautioned, "you boys don't want that fellow on your force."

"Why, Mother?" I asked.

"Because he is no good," she said.

"Why Mother, I have always found him to be a pretty good fellow," I replied.

"Yes, but Mooney, the jails, penitentiaries, insane asylums, and hell are all full of good fellows and you and I and everyone else are a damned sight better off because they are there," she said, laughing. Changing the subject, she inquired, "What are you going to do tomorrow?"

I looked at her and grinned rather sheepishly. "I am thinking seriously of climbing Popocatepetl," I told her.

"You'll never leave here satisfied unless you get to climb that damned mountain, will you? Well, I guess it's all right when you are able to do those things, but I am not able to do them any more, Mooney. I have been to the top of the hill and am now approaching the valley for the last time," she mused.

"Bosh! Mother," I said. "If you continue to live the exciting life you have had up to now, you will live to be a centenarian."

"I don't know, Mooney," she said, looking rather wistful, "I am willing to stay here that long if I can do any good for the wretches and do not become a burden on someone in my last days. Well, go to bed now and get some rest for you will need it if you are going to climb that mountain tomorrow."

The next morning I arranged to be taken to the pack-horse station on Popocatepetl. I rented a horse and eventually realized my ambition. Climbing into high altitudes has always been a hobby with me. My liking for this pastime is based upon the belief that when up on top of peaks I am near the first mover of all things that move.

Two days after my trip up Popocatepetl I was regretfully compelled to terminate my stay in Mexico City. Before leaving I went to Mother Jones' room to bid her good-by. While there, a messenger arrived bringing a package from President Obregón to her. Mother opened the package after the messenger had departed and disclosed its contents. In it was a note which read, "With love, respect and devotion," from Alvero Obregón. The package contained 25 American double eagles in gold ($500.00). "Truly," I said to her, "nothing in Mexico is too good for 'Madre Yones.'"

During our conversations with hundreds of the Mexican people, we learned that Ambassador Morrow was loved and respected by every one. We had planned to visit the U. S. Ambassador but had not gotten around to this event when I had to leave. Mother told me upon her return to the States that she had enjoyed a very pleasant visit with Ambassador Morrow before her departure from Mexico City.[7]

## Chapter VIII

Early in 1921, a representative of the Associated Press was en route from Bluefield to Kenova via the Norfolk & Western Railroad. During that trip he engaged a prominent coal operator (then operating in that field) in conversation.

"We are not going to fight the union like we have been fighting it," the operator confided. "We have been spending fabulous sums to hire guards to keep the organizers out, but we are not going to do that any longer. We are going to secure indictments against the officers and active members of the union, throw them into jail, deny bail, prosecute them, and even if we do not secure many convictions we will keep the union bled of funds and break them that way."

In this way he foretold the approaching climax of a conspiracy between the law and the reign of unlawful violence that had been operating in that part of West Virginia for decades. How well this conspiracy operated, with the assistance of governors, county sheriffs, judges, state police, and appointees of the governor until the second intervention of federal troops, will be unfolded as we proceed.

Soon after the above conversation occurred Mingo County was plastered with a declaration of martial law, and scores of miners and several officials were thrown into jail and denied bail. We branded the declaration as "pseudo" or "bunk", and appealed to the Supreme Court for the release of three men on a writ of habeas corpus. The court decided that "in the absence of a military force to effectively maintain a state of martial law, the proclamation was ineffective and did not obtain."

The military force used to maintain the governor's[1] proclamation, until the Supreme Court ruling, was composed of riffraff not only of Mingo County but from other parts of West Virginia. Many of them were from some of the adjoining states. Not to be outdone (and ever subject to his master's voice), when his first attempt to break the spirit of the strikers failed, the governor re-organized the militia and re-declared martial law.

During the summer of 1921, we initiated a move to have a committee of the United States Senate conduct an investigation into the industrial conditions then prevalent in West Virginia, or that part of the state affected by the strike. We were unsuccessful in our first attempt and the resolution was tabled, but I eventually persuaded Senator Hiram Johnson of California to re-introduce the resolution and it passed the Senate.

During the hearings before a sub-committee of the Senate Committee on Education and Labor, we asked Sid Hatfield to appear as a witness. While in Washington, he was notified by wire that he had been indicted in McDowell County which adjoins Mingo.[2] I accompanied him to the depot at Washington when he left for home. His parting words to me were as follows: "Well, old boy, I will never see you again."

"Don't feel that way about it," I admonished him, "you will come out all right."

"No," he said, "Felts knows that I have not been in that county and he framed this indictment to get me up there where they can kill me and Ed."

Sid Hatfield, Ed Chambers, and John Collins were arraigned and gave bond for their appearance in answer to the indictments; trial was arranged for a later date, August 1, 1921.

Mr. C. J. VanFleet of Charleston was dispatched to conduct the defense of the three defendants at Welch, McDowell County.

Hatfield and Chambers were accompanied by their wives when they appeared for trial. Upon leaving the train at Welch they went to a hotel and in compliance with the request of Mr. VanFleet, Hatfield and Chambers left their guns at the hotel. Mr. VanFleet preceded them to the courthouse and as they approached the steps leading into the main building there were in excess of 40 gunmen sitting on the balcony or standing at vantage points.

Sid and Mrs. Hatfield were in the lead; following them were Ed Chambers and his wife. Bringing up the rear was the other defendant, John Collins, accompanied by James Kirkpatrick.[3] Kirkpatrick had been requested to go along as a friend. Hatfield and his wife had mounted to the second or third step and the

Chambers were just at the bottom of the flight when from in front and on both sides a deadly fusillade of shots was fired into the bodies of Hatfield and Chambers.

Hatfield wilted first under their withering fire, and Mrs. Hatfield turned to run and fell in a dead faint. Not so with Sally Chambers; when Ed began to sink, she grabbed him as though to shield his body from the leaden hail, calling out to him yet fully realizing that he would never again answer her in this world. Turning on C. E. Lively, who was sending bullet after bullet into the body of her husband, she fought him with her parasol.

Kirkpatrick and Collins ran for the train and were not missed until the excitement had somewhat subsided. The conductor locked them in a compartment on the train and refused to reveal their hiding place to the gunmen who searched the train for them.[4]

An eye witness told me that one gunman by the name of Nathan Ash emptied his Colt into the bodies of these men and never ceased puffing on his cigar. Seventeen bullets penetrated Hatfield's body, while Chambers' body was the recipient of 13. Neither body could be embalmed.

The funeral of these two men was attended by hundreds, and Samuel B. Montgomery, now deceased, of Kingwood, was elected to deliver the oration. He was grand keeper of records and seals for the Knights of Pythias of West Virginia, and both Sid Hatfield and Ed Chambers were members of that order. Amid a steady downpour of rain which continued throughout the funeral proceedings, he said in part:

"We have gathered here today to perform the last sad rites for these two boys who fell victims to one of the most contemptible systems that has ever been known to exist in the history of the so-called civilized world. Deliberately shot down, murdered in cold blood, while they were entering a place which should have been a temple of justice, and by whom? Men who are working under the direction of and taking their orders from coal operators who live in Cincinnati, Chicago, New York City, and Boston.

FRED MOONEY AND FAMILY, *circa* 1934, at Wendel, Taylor County, the location of Maryland Coal Company operations.

LEADERS OF DISTRICT 17, *circa* 1917. Left to right: William Blizzard, president of Sub-district No. 2; Fred Mooney, secretary-treasurer; William Petry, vice-president; and C. Frank Keeney, president.

LEADERS AND REPRESENTATIVES OF DISTRICT 17, UNITED MINE WORKERS, in front of the Jefferson County Courthouse, Charles Town, at the time of the treason trials in 1922. Left to right: James M. Mason, Jr., C. J. Van Fleet, Samuel B. Montgomery, and Harold W. Houston, lawyers; William Blizzard, president of Sub-district No. 2, St. Albans; John L. Lewis; and C. Frank Keeney. Men in back row and those on porch not identified.

ALBERT "SID" HATFIELD, 1919.

"Sleek, dignified, church-going gentlemen who would rather pay fabulous sums to their hired gunmen to kill and slay men for joining a union than to pay like or lesser amounts to the men who delve into the subterranean depths of the earth and produce their wealth for them. At the same time these same men prate of their charities, their donations to philanthropic movements, act as vestrymen and pillars of the churches to which they belong.

"Even the Heavens weep with the grief-stricken relatives and the bereaved friends of these two boys."

Neither Keeney, myself, nor any of the other officers could attend that funeral because of martial law. When Mr. Montgomery returned to Charleston, several hundred miners were congregated at the miners' hall on Summers Street and they insisted on having a meeting. He spoke to them for a few moments and they went their several ways to their homes. Mass meetings were held in many communities, petitions were signed, letters of protest sent to state officials, etc., but to no avail.

The killing of Hatfield and Chambers occurred on August 1, 1921. Two or three days later Mother Jones came to Keeney and me and asked us to call a mass meeting for her somewhere in the Kanawha Valley. We advised against such a move owing to the tense conditions that prevailed following the death of Chambers and Hatfield. When she insisted that a meeting be called, we flatly refused to comply with her request. Following our refusal she became abusive, and going into the assembly room where 25 or 30 miners were congregated, she proceeded to read our pedigrees in true Mother Jones style. She told them that "Keeney and Mooney have lost their nerve; they are spineless and someone must do something to protect these miners."

Our idea of protection at that time was to let sentiment cool down and reach a normal state. On August 7, a mass meeting of 700 or 800 miners did congregate on the old Capitol site at Charleston. Mother Jones, S. B. Montgomery, and myself spoke to them for short intervals. But we talked in a conciliatory strain, advising them to go home, keep at work, and let the law take its course. The meeting disbanded and the next day Mother jumped all over William Blizzard and accused him of calling that mass meeting. Hot words ensued between the two and Mother went away.

By August 15 or 16, men were moving in the general direction of Marmet. A group of men would bivouac there for two or three days and then scatter; others would drift in and out. Mines were being closed down in violation of contract, men drifting around, milling from place to place. We were worn out, caught between the restlessness of the miners and the insistence of the coal operators that we keep the mines in operation.

This condition prevailed until the last 10 days of August. About August 24, the miners began gathering in earnest and by the 26th approximately 12,000 were under arms and were either in camp on Lens Creek near Hernshaw or were moving in the general direction of Logan County by the way of Madison and Danville in Boone County. On August 24, a committee from the miners camped above Hernshaw came into the office with the following information and request:

"Mother Jones sent word that she would be up this afternoon to speak to us and that she had a telegram from President Harding which she wanted us to hear. We want you and Keeney to come to that meeting and verify that telegram and explain why (if you know) President Harding did not send the telegram to the officers of the union instead of sending it to Mother Jones?"

Keeney and I went to that gathering, and Mother was speaking when we arrived. There were only 350 or 400 men present. The main body of the men had moved on across the mountain towards Madison. Mother read the telegram after we arrived. It read about as follows:

"To the miners encamped at or near Marmet with the avowed intention of marching on Logan and Mingo counties. I request you abandon your purpose and return to your homes and I assure you that my good offices will be used to forever eliminate the gunman system from the state of West Virginia. Signed, Warren G. Harding, President of this great republic."

The assembly demanded of us that we verify that telegram so Keeney asked Mother if he could see the copy she read. "Go to hell," she said, "none of your damned business." Not knowing that Keeney had been rebuffed in the aforementioned manner and having become separated from him in the crowd, I asked Mother if I could see the telegram.

"No you cannot, none of your business."

Men began to mill around, saying, "Let's go home. Harding has promised to remove the gunmen." Others cried out, "That damn telegram is a fake or she would have shown it to some one."

Some one below where we were standing called out, "Throw a guard across that road and don't let anyone through until we find out about this."

They gathered around me and insisted that I take a committee back to Charleston and verify or refute that telegram. Upon our return to the office I wired President Harding as follows: "Telegram read to miners here today purported to be from you pledging federal aid in ridding West Virginia of the gunman system. Will you confirm that telegram through this office."

A few hours later the following telegram was received from George B. Christian, secretary to the President: "President out of city. No such wire sent by him." We told the committee that the advice given in the telegram was good even though it was unauthentic insofar as coming from the President was concerned. But these men were mad, everyone was mad, seething to revenge fallen friends, and they were total strangers to reason.[5]

Additional hundreds of marchers kept pouring into Lens Creek by way of Marmet and the columns moved toward their avowed destination.

About 5:00 a. m. on August 27, Brigadier General H. H. Bandholtz called Keeney and me out of bed, commanding us to meet him at the governor's office at once. We called Attorney Houston and asked him to accompany us.

The general never wasted any words. He said as I recall, "You two are the officers of this organization, and these are your people. I am going to give you a chance to save them, and if you cannot turn them back we are going to snuff this out just like that," and he snapped his fingers under our noses. He further said: "This will never do, there are several million unemployed in this country now and this thing might assume proportions that would be difficult to handle."

We told him that we believed it would be useless, but we would make an effort. "Would you arm us with a statement signed by you telling the miners what the consequences will be if they do not turn back?" I asked the general.

"No," he snapped.

"Then I believe our trip will be fruitless," I insisted.

"Yes, I will give you a note," he said, and immediately he dictated one, had it transcribed, and handed it to us.

Securing a taxi, we picked up W. T. Dadisman, secretary of Sub-district No. 3 who was stopping at the Lincoln Hotel, and started out to intercept the advance columns. While en route up Lens Creek, we passed several small groups of miners encamped or moving on to join the main body.

Just above Hernshaw, which is located four miles from Marmet, we passed about 80 miners who were armed only with pistols and revolvers. We broke the news to them and moved on, but as we were leaving them they fired several hundred shots into the atmosphere. Five or six hundred yards further on and just around a bend, we approached a truck load of provisions standing in the road. No one was with the truck or in sight anywhere. Instantly it occurred to me that those with the truck had heard the shooting below and had gone into ambush.

So I called out, telling them who we were and about 12 men came out from behind trees, rocks, and hummocks letting down the hammers of their rifles. "Boys, boys," they cried, "in two more seconds we would have fired on you."

The first large group of marchers was encountered near Peytona, and after spending sufficient time to acquaint them with developments we moved on along the intercounty highway which led up Droddy Creek,⁶ thence across the mountain into Danville. Marchers were strung out along this creek from mouth to head. Each company was under the direction of an ex-soldier. Seven hundred of these boys were in the march and some of them were veterans of both the Spanish-American War and the World War. In charge of one of these companies was Harvey

Dillon, of Winifrede, a veteran of both wars. When he saw who was in the cab he split his company, forming an aisle through which we passed. He stopped at the advance end and blocked the way:

"Boys, what are you fellows doing here?" he asked. After reading General Bandholtz's note to him, we explained the situation. "Boys, are you telling us straight," he demanded.

"Read the ultimatum," I said, handing him the note.

"That's his signature all right for I was an orderly under him in the Philippines and I would know his signature anywhere." Turning to his company, he said: "Salute! damn you salute." Following the salute he turned to his men. "Boys, we can't fight Uncle Sam, you know that as well as I do." Turning to us he inquired, "What shall we do, turn back from here?"

"No," we advised. "Take your men on into Danville and wait until we can arrange for special trains to take you out."

During the entire conversation just recorded, Dillon was crying as were many of his men. Tears of defeat coursed their way down many cheeks. The majority of his men were dressed in regulation army clothing, including helmets, and several of them were carrying gas masks.

We overtook the advance columns as they were leaving Madison, pushing on in the direction of Blair in Logan County. We prevailed upon them to turn back and congregate in the ball park at Danville until we could talk to them. "We are going on," they insisted.

"Boys, you better listen to reason, for you are all going to be slaughtered if you proceed," we warned.

When they were convinced that we were really telling the truth, they agreed to return to the ball park and go into camp with those with whom we had conferred along the route.

After securing a room and some food at a hotel operated by Mr. P. C. Dunlap, we followed the remaining marchers to the ball park at Danville. During the meeting, many of the miners and their friends were bitter and denunciative toward us because we had intercepted them; but after we read the message from General Bandholtz they quieted down and became more reasonable. However, even then we were skeptical about permitting them to take a vote. "We are not going to call for a vote, we are just going to ask you to take our advice and let us take you out of here," Keeney admonished. Many grumbles and threats were heard, some cried, some cursed, while others acted as though they appreciated the effort we made to get the message to them on time.

We returned to the hotel and I tried to arrange for special trains to transport them out of the territory.

Sundown brought Lewis White from Blair. Lewis White was a brother to Deputy Sheriff White of Logan County, who acted as jailor during the administration of Sheriff Chafin. He was riding a velocipede, was coatless, and carried two Smith and Wesson revolvers of the latest type. When he stopped at the depot where Keeney and I were standing he said, "What the hell you fellows mean by stopping these marchers."

"To prevent them being slaughtered," we informed him.

"Oh, hell! What you two need is a bullet between each of your eyes," he said. Being worn out and not in very good mental or physical condition, it was "nip and tuck" which one of us took the lead in extending him an invitation to try his hand.

From Lewis White had come many of the reports that the gunmen were firing on and had killed women and children at Blair. I believed then and believe now that White was an undercover operative for the gunmen.

Fifteen or 20 minutes after Lewis White arrived at Madison the commandeered train came down, and after he boarded the coach it went on to the ball park at Danville. Speeches were made to the miners by White and several of his colleagues in an effort to get them to board the train and go on to Blair but to no avail.

It was several hours before any promise of trains could be secured and eventually it appeared that we were being thwarted through the governor's office and that this situation must be overcome. There was an Associated Press representative along with us and he said, "Let me get on that phone and see what I can do." He went through to Pittsburgh, contacting his superiors by telephone, and soon afterward we were notified that the trains were on their way and would arrive about 3:00 a. m.

The trains arrived at 5:00 a. m. and the miners were crowded aboard. These trains were continued from St. Albans to Fayetteville and way points. Several extra street cars were run to Cabin Creek Junction. By midnight, August 29, the miners were all at home and peace and quiet reigned.

General Bandholtz returned to Washington, satisfied with the results obtained.

August 30, 1921, was a beautiful day on Coal River. The mining towns of Sharples and Blair were quiet. Small groups of miners stood here and there discussing the previous hectic days or went their several ways slowly but surely drifting back to normal routine. The day passed and nightfall saw small groups around outdoor fires still engaged in the process of talking themselves out. One such group was standing around a small fire near the coal tipple at Sharples when from the darkness came the command to "throw up your hands you red necks." These miners were not permitted to surrender but with the command a fusillade of shots was fired into the group killing three and seriously wounding two more.

Three hundred and fifty deputy sheriffs, gunmen, and state police,[7] under the leadership of Captain J. R. Brockus of the state police, had sneaked across the mountain from Logan under the pretext of serving warrants on someone at Sharples and committed what General Bandholtz termed "An unwarranted and untimely attempt to make arrests," thereby precipitating the second march on Logan County.

News of this act traveled like a forest fire. Men began to mobilize from all the surrounding territory. Coal company stores were looted for guns, ammunition, and supplies. Trucks, automobiles, teams, and trains were confiscated. Anything and everything that meant transportation was commandeered at the point of guns, loaded with men, and started for Logan County.

There were no bivouacs anywhere; if a vehicle broke down and was not easily repaired it was discarded and another one picked up the stranded and rushed them on to Blair. Blair Mountain was the main gateway to Logan when traveling from Charleston through Hernshaw, Danville, and Madison.

This area was the scene of one of the most bitter struggles recorded in the history of industrial warfare. The fight began on the morning of August 31, 1921; the contending forces were halted by the appearance of federal troops on September 6.

The intercounty highway from Charleston, through points previously mentioned, crossed the railroad tracks at the southern end of Blair, followed a ravine to the top of Blair Mountain, and passed through Blair gap and down Dingess Run to intersect with the Guyandotte River highway at Stollings, three miles from Logan, the seat of Logan County.

The C. & O. Railroad from Madison to Blair was commandeered by a group of men under the leadership of Lewis White. Midday August 31, 1921, there were between 7000 and 8000 men in the vicinity of Blair, and several skirmishes had taken place. Hundreds of these men were transported to Madison by auto and truck, then taken aboard the commandeered train and rushed to Blair via railway.

Rev. J. E. Wilburn was the father of a large family among whom were two sons, John and Isaac. Prior to September 1, the head of the Wilburn family steadfastly refused to take part in the activities of the miners. "I am, or was, a God-fearing man," he told the miners, "until one morning while sitting at my breakfast table I started to pass a dish of food to one of my boys and a bullet fired from Blair Mountain came through the house and broke the dish in my hand." He arose and walked to the front door and out on the front porch. He called to several of the assembled miners and asked them to gather around for he had something to say to them.

"Boys," he said, "I have refused up until this time to have anything to do with or take part in your activities, but this morning I am laying aside my Bible until this community is made a safe place in which to live. I am ready to go to the front and when I get there the gunmen are going to know I am in action."

He was indicted, tried, and convicted for the killing of Deputy Sheriff John Gore of Logan County. John Wilburn, son of the Rev. J. E. Wilburn, was also indicted, tried, and convicted for being an accessory to the killing of John Gore.

Sheriff Don Chafin was King in Logan County. It was obvious to us that even Judge Robert Bland of the criminal court of Logan County made no move whatsoever without consulting Sheriff Chafin. Operating in Logan County were several hundred deputy sheriffs deputized by Chafin and paid by the coal companies. It was admitted under oath that during a period of six months in 1921, $46,000 was paid to these special officers by the coal companies in Logan County.

The purchase of a railroad ticket at Huntington for any point on the Guyandotte branch of the C. & O. Railroad was sufficient cause to be investigated by the gunmen. The business of anyone who was not known to the gunmen was inquired into, and if answers were not satisfactory, they would be held under guard at Logan until the next train returned to Huntington and escorted aboard with the admonition to never return.

Several instances are recorded where traveling men were assaulted, beaten up, and deported out of Logan County. Among these were one or two cases where the assaulted parties brought suit in the federal courts (because they resided in another state) and damages were sought. In some of these cases the authorities of Logan County denied ever seeing the defendants or having knowledge of the aggrieved parties being in their bailiwick.

When the miners and their sympathizers began to congregate at Blair, Sheriff Chafin was quoted as saying, "Let them come on, Logan County is ready." Sheriff Chafin had called for volunteers not only from Logan County but from Mingo and McDowell counties as well. Both factions of the contending forces were well armed, and based upon evidence produced in later trials, the only advantage Don Chafin's side had was the possession of one or two planes from which crude bombs were dropped in an attempt to annihilate the miners.

Both factions were equipped with modern machine guns and up-to-date rifles. It was alleged by the prosecution that a Browning water-cooled machine gun, that had been stolen from the Express Company after it had been received for delivery to a coal company at Weirwood, was used in action against the forces of Sheriff Chafin in Logan County. However, this allegation was never definitely proven.

The bombs used by the Logan County authorities were constructed of two sections of gas and steam pipe; one section being smaller was inserted into the other. The space between was filled with scrap iron, bolt-nuts, and other cuttings from a machine shop. The inner pipe was charged with seven or eight pounds of gunpowder with a 12 gauge shotgun shell, minus the shot, set into one end as primer. These bombs were dropped in several places during the fighting but no casualties were reported.

Fulton Mitchell and three deputy sheriffs were captured in an old abandoned hut on Hughes Creek and held prisoners for several days. Many persons advocated the execution of these prisoners, but through the intervention of someone reason prevailed and they were released when the federal troops arrived.

General Bandholtz was again rushed to the scene with detachments of troops from Camp Dix and Camp Taylor. Upon the arrival of the federals the miners again surrendered their arms and demanded protection of them.

Three or four casualties were reported by both factions, but the major damage was done to rocks, trees and shrubbery. Witnesses later swore that William Blizzard appeared at Blair during the battle and delivered an automobile load of ammunition and while there made a speech to the miners encouraging them to continue the fight. However, this testimony was not sustained by the jury when Blizzard was tried for treason. Nevertheless, he was referred to by the press as "Generalissimo Blizzard" of the miners' army. Concrete dug-outs were installed by the Logan County authorities and it was reported that some of these were captured by the miners.

Sunday night, August 31, a carload of men came to my front gate and called me. Going out, I asked them what they wanted.

"Listen," said one of them. "We think more of you and Keeney than we do of any other two men on earth, but we are going to Logan County this time and if you don't stay out of our way we are going to treat you just like we intend to treat the gunmen when we meet them. The best thing for you two to do is to clear out and stay out until we get through here. Get clear out and stay out until we get this thing cleared up."

I immediately drove to Keeney's residence on Edgewood Avenue and related the news to him. We had received notice a few hours before of the indictments that were to be returned against us in Mingo County. These charges included five counts ranging from misdemeanor to murder.

After thoroughly discussing the situation we reasoned that if we were incarcerated while the excitement was at fever heat that we would in all probability be treated as Sid Hatfield and Ed Chambers had been served when they appeared for trial at Welch on August 1. Viewing the situation from all angles we decided to "clear out" for a few days at least.

Two days prior to the events just recorded, James Riley, then president of the West Virginia State Federation of Labor, Frank W. Snyder, editor of the *West Virginia Federationist*, H. W. Houston, chief counsel of District 17, United Mine Workers, and myself met at the office of the United Mine Workers and organized the Mingo County Defense League. It was agreed that I should act as treasurer and secretary of the league. Circulars were mimeographed and sent to all parts of the world soliciting funds for legal defense.

This fund eventually totaled $46,000.00, which was used to defend those who were tried from among the 1,276 who were indicted for alleged participation in the "March on Logan and Mingo County." I shall have more to say of this defense fund.

Leaving Charleston at 12:15 a. m. September 1, Keeney and I crossed the Ohio River to Pt. Pleasant. R. M. Williams, then a district field worker, transported us to Columbus where with the assistance of John "Jock" Moore we found suitable headquarters on Long Street just off High.

Strange as it may seem, we later learned that Don Chafin was quartered at the Neil House just a few blocks away at the same time we were on Long Street. We kept in touch with developments through the press reports and by messenger until September 16. On this date we met with counsel at Athens, Ohio, and decided to return to Charleston and surrender to the authorities. With the assistance of Samuel B. Montgomery and J. H.

(Peggy) Gibbs of Hartford City, we returned to Charleston and hid away near our homes until some of the acting officers of the union could arrange a meeting with the governor.

At 5:00 p. m., September 18, we appeared at the governor's office and were taken in charge by deputy sheriffs John Copenhaver and T. Newcomer of Kanawha County on the warrants issued in Mingo County. They delivered us to the authorities at Williamson.[8]

## Chapter IX

Keeney and I were incarcerated in the Mingo County jail in Williamson on the night of September 18, 1921, and held there until November 11. When we arrived as guests of Sheriff Pinson we were placed in a suite of cells along with board member C. H. Workman, field worker A. D. Lavinder, James Kirkpatrick, Lawyer Cline, Pleas Chaffin, A. H. Boggs, Ace McCoy, and others whose names I do not remember. These men had been confined in jail by Major T. B. Davis, then in charge of the enforcement of Governor E. F. Morgan's pseudo declaration of martial law in Mingo County.

Keeney and I were kept in cells only a few days when the jailor (Mr. Messer) offered us an upstairs room in his house which was built in connection with the jail. We occupied this room until our wives came to visit us, at which time he gave us a room each during their stay. Judge Bailey admitted us to bail immediately upon our arrival in Mingo County, but we would not furnish bail because of the capias warrants which had been sent there from Logan County.

Rumors were constantly afloat that the gunmen were going to raid the jail and kill all the miners who were inside, so we had a friendly striker smuggle a pair of six shooters inside to us and we slept with them under our heads at night and kept them well in reach in daytime. Signals were arranged with some of our friends on the outside so we would be warned if an attempt was made to storm the jail.

One night we were awakened by several shots and a commotion directly under the window of our room. Grabbing our guns, each of us ran to a window and cautiously looked out. A hatless militiaman was cursing the clothesline while rubbing his nose. Someone had broken into a store uptown and while chasing the robber, this would-be soldier had run into the clothesline. Thinking the enemy was upon him, he proceeded to shoot the atmosphere full of holes.

Mr. Messer was up and outside before the soldier fully recovered from the attack. "What is coming off here," he inquired.

"Was chasing a robber and ran into that damn clothesline. It near busted my nose and caused me to fire my gun when I fell," he replied.

The jailor came up to see if we were in on the excitement. "Did you hear and see the fun?" he asked.

"Yes," Keeney replied, "when the bombardment began we got out of bed." We enjoyed a laugh at the expense of the law.

One evening word came into the jail that Don Chafin was in town and was coming up to see us. To say the gang was frightened would be putting it mildly. We were scared stiff. Someone called up the stairs and said, "Boys, here comes Chafin."

Chafin entered the jail with a man named St. Clair who posed as a mine inspector. When Chafin reached the top of the first flight of steps he was facing the group of cells which held Ace McCoy and Pleas Chaffin. McCoy called through the bars, "Well is that the bad man, Don Chafin? Why he don't look bad, does he?" Chafin laughed and as he approached the cell front where we were, through habit or otherwise, he placed both hands behind him in the neighborhood of his hip pockets.

"Hello, Keeney, which one are you?" he asked. He shook hands with Keeney, Workman, and several others in turn. A. D. Lavinder and I remained in the cell.

"Where's Lavinder?" Chafin asked.

"Here I am," Lavinder replied, but he did not get up off the bed to shake hands with Chafin. Chafin did not ask for me and I was glad he refrained from doing so for when the call came up the stairs that he was on his way in, I told the gang to stay close to the floor if anything started and I sat down in the back of the cell with a Colt ready for instant use.

Mr. Workman was twitted for being the worst frightened man in the group after Chafin and St. Clair had gone, but I believe that if temperatures had been taken and heartbeats registered, the kettle would not have been able to say to the pot, "you are black."

The next day a resolution was introduced on the floor of the mine workers' convention, then in session in Indianapolis, providing for a committee to be appointed to wait upon President Harding and demand protection for the several hundred coal miners and their officers who were confined in the jails of Logan and Mingo counties in West Virginia. This committee was instructed to get in touch with John "Jock" Moore of Ohio, who was then legislative representative of the United Mine Workers at Washington, and enlist his aid in obtaining an interview with the President.

Jock Moore told me that when President Harding was made acquainted with the situation he said, "Why, Jock, they would not murder men who were being held for trial, would they?"

"Mr. President, they have done so," Moore replied.

Jock also informed me that President Harding called Governor E. F. Morgan of West Virginia over long distance telephone and told the governor that if any more men were shot or mistreated while awaiting trial he would hold him responsible for the acts.

The suggestion was made to President Harding that he keep in touch with the situation in West Virginia and obtain first-hand information regarding the treatment of prisoners confined in the two county jails. Two or three days later several operatives from the intelligence department came into Williamson and one Captain O'Bryan sent a note into the jail addressed to Keeney and me as follows:

"Would come in to see you but think it best not to do so. Rest assured that the situation is being well cared for."

We saw Captain O'Bryan and one or two of his associates in the neighborhood of the jail on several occasions.

Up to this time Mr. H. W. Houston of Charleston had been chief counsel for the miners of District No. 17. He had handled many important cases including the trial of the defendants who were charged with killing the Baldwin-Felts detectives at Matewan in 1920. However, there were two counties involved which Houston could not enter. Logan and McDowell counties had barred him from practice on penalty of death. Mr. C. J. Van-Fleet, who was there to defend and who was inside the courthouse at Welch when Sid Hatfield and Ed Chambers were killed, was told on several occasions: "Don't let that lawyer Houston come up here. If you do we will kill him." This threat was made by the gunmen. The same warning was given out on several occasions in Logan County, only in a different way. The gunmen were quoted by many as saying, "Bring that fellow Houston up here, we want to get at him anyway."

Following a conference, Keeney and I decided to secure the services of T. C. Townsend of Charleston. Townsend was sent for and when he had thoroughly discussed the legal aspects of all the pending cases, he suggested that we apply for changes of venue and try to have the cases tried in some distant and remote county. We knew that the Logan County capias warrants were in the hands of the sheriff of Mingo County and that if we gave bond we would be railroaded into Logan. Townsend went out to the courthouse and came back with his face creased in smiles. "Boys, those capias warrants have run out. The Mingo authorities cannot hold you for the Logan authorities on those warrants. They have been dead several days."

Having been admitted to bail in Mingo, we decided to arrange for two deputies from Kanawha County to be on hand on November 11, and when we furnished bail in Mingo County they could take us in charge upon presentation of capias warrants from Kanawha County where indictments had also been returned. We reasoned along with Townsend that if we were given a trial in Kanawha County, there would be no possibility of an assault, pending or during a trial, and that we would at least get a fair trial, which was unknown in Logan County when a union man or a union sympathizer was concerned.

On November 11, two deputy sheriffs from Kanawha County arrived in Williamson, Pike Trobridge and T. Newcomer. Trobridge was a veteran officer, well-liked and respected by everyone. Newcomer also had legions of friends in the territory where he operated as deputy sheriff under H. A. Walker of Kanawha County. Sheriff Pinson of Mingo County was in Logan County vacationing with Don Chafin when Trobridge and Newcomer arrived. John Hall, chief deputy under Pinson, was in charge of the office when we walked into the courthouse to give bond.

"We wish to furnish bail for our release," Hall was informed.

"All right, boys," he said, "but I suppose you know that the capias warrants from Logan are here and as soon as you give bond we are going to hold you for the Logan County authorities."

"You cannot hold a man on a warrant that is dead," Townsend warned him, "and the warrants from Logan have been dead for seven days."

"Dead or alive," Hall shot back, "I'll hold these two men on these warrants from Logan."

"If you try it, you will do so at your peril," Townsend snapped.

"Wait until I call the sheriff and see what he says about it," Hall requested.

Following the telephone conversation with Sheriff Pinson, Hall announced that we were free to go and that the Logan authorities would get us when they wanted us anyway.

Several of the strikers came to us and reported that 40 or 50 gunmen were gathered in the vicinity of the Norfolk and Western depot and that we had better be careful.

For the benefit of those who are not acquainted with the town of Williamson, a limited description might not be amiss. The main line of the Norfolk and Western Railroad, running north to the Great Lakes and south to Roanoke, Virginia, passes through Williamson. The town has a population of approximately 7,000. It is bound on the west side by Tug River and on the east by the mountains. Tug River also constitutes the boundary line between the states of West Virginia and Kentucky. The principal business section of Williamson, including the courthouse, county jail, hotels, general merchandise stores and banks, is located between the railroad and Tug River.

Because of the railroad passing through town on a low grade, underground tunnels and crossings were made, one above or south of the depot for pedestrians only and one below or north of the depot for traffic. From the east end of the tunnel for pedestrians to the depot platform where passengers are compelled to go before trains can be reached would be a distance of approximately 250 feet.

When the news went forth that Keeney and I had given bond and would in all probability be taken away on the midnight train, about 40 gunmen took up positions on either side of the route from the east end of the tunnel to the depot platform. If we had traveled this route in going to the depot, I think we would have been shot down. Not an atom of doubt has ever entered my mind since that night but what that would have been our last walk. However, not without giving an account of ourselves, for we were prepared for and expecting the worst.

But we did the unexpected.

The question was asked in the courthouse, "Are you folks going out tonight?"

"No," I replied, "We are not going out until tomorrow."

We arranged for eight men in addition to the two deputies, Keeney and me, to make the trip to Charleston. The eight were tried and true and we knew they would stay to see the finish in any emergency.

"Have two closed cabs behind the jail about 10 minutes before train time," we instructed the eight men, "and you fellows be in one of them." At the allotted time the four of us walked out through the back door of the jail and entered the waiting cabs.

We drove to the north traffic crossing and with dim lights awaited the arrival of the train. When the train pulled into the depot, we had the cab drivers edge up near the combination baggage and smoking car. Before the gunmen knew what was happening we were in the combination car, blinds down, backs to the wall.

Several strikers mingled with the gunmen to note their chagrin when they learned that they had been cheated of their slaughter. One gunman by the name of St. Clair, previously referred to in connection with the visit of Don Chafin, went mad. A striker described his state as "frothing at the mouth, pulling his hair, and cursing everything and everybody."

After we were seated in the car two of these professional killers had the unmitigated gall to come aboard and look us over. Our trip to Charleston was otherwise uneventful and we were turned into the basement bull pen of the Kanawha County jail.

Jails are inhuman institutions. I saw strong men become nervous wrecks; men who were believed to be normal broke down, cried, and sniveled like babes.

Two or three days after our entry into Kanawha County jail William Petry, then vice president of District No. 17, and William Blizzard, president of Sub-district No. 2, came and surrendered to the Kanawha County authorities. Shortly after Petry and Blizzard came into the Kanawha County jail we simulated the "kangaroo court." Keeney took the position of judge, Petry was prosecutor, Blizzard acted as sheriff, and I assumed the role of defense counsel when prisoners were arraigned for breaking into jail without permission of the inmates.

But the "cat o' nine tails" was not present on the floors where we maintained jurisdiction. While we were in charge of it, the "kangaroo court" was conducted in satire.

During our stay in Kanawha County we were interviewed by several newspaper reporters and asked to give out statements for publication. There is one thing concerning which any individual can rest assured. If the subsidized press will not print your views, get into jail or in the clutches of the law and then the press agents will swarm around you like hungry sunfish waiting for a worm. It matters not what you have to say or how you say it, they will broadcast your views to the world.

One reporter asked me my views concerning the cause of the predicament in which we found ourselves at that time.

"We are caught between the nether millstones of a gigantic conspiracy hatched between the law and unlawful violence," I replied. "Those who are trusted with the enforcement of existing law know and admit that the gunman system against which we have been fighting for 20 years is wrong and the workers along with their trade union representatives are victimized in order to cover up and camouflage official cowardice.

"Community political support is bargained for by each politician. Each community has its one or more hobbies or little evils, as they are often called. In one community or county it is the system of employing gunmen under the pretense of protecting private property but in reality and practice to keep employees in subjection. In another it may be gambling or moonshining. In still another it may be red-light districts. Ever and anon it is the same. No matter what the evil or hobby, he must promise to pursue a policy of non-interference in return for support. When he is elected to office he must regulate his official acts to conform with his promises.

"Politicians must be obscure. The chief prerequisite of an 'A No. 1' politician is obscurity. He must be capable of apparently saying much and yet saying nothing. In order to receive the backing of the party he must be able to couch his speeches and interviews in such terms that when called to account he can say, 'Oh, well, I did not mean that, I meant this.'"

Political speeches often remind me of an incident which occurred during one of Sam Jones' sermons. He extended an invitation to a congregation of several thousand people as follows:

"All the married couples in the congregation who have been married for a period of 10 years or more and have never had a misunderstanding, an argument, or a fight, please stand."

About 25 couples stood up. Then Jones requested that the remainder of the congregation kneel and pray for these standing liars.

During our incarceration, Governor E. F. Morgan of West Virginia was invited to address the state bankers association. He said in part that "Keeney and Mooney surrendered to me at the statehouse and they had the temerity to demand protection." We did demand protection from the brand of violence that had snuffed out the lives of Sid Hatfield, Ed Chambers, and another man who was killed inside the Logan jail while the armed march was in progress.

Kanawha County was home to Keeney, Blizzard, and me. We had been reared within its confines. We felt that we were among friends insofar as politicians can be friendly to men who represent labor.

However, the Kanawha County jail was one of the worst holes I had ever seen. It reeked with filth and vermin. Many times while we were there the sewage backed up and stood shoe-mouth deep over the entire basement. At night damp-footed sewer rats often disturbed our slumber by clawing across our faces or running the full length of the bed. Approximately 200 prisoners were confined in that jail and the normal capacity was about 100.

About 40 of the inmates were Negroes, and they were confined in the bull pen on the ground floor. "Kangaroo court" was in full blast among the "shines" and men were beaten unmercifully for trivial offenses against the assumed peace and dignity of the inmates.

The "whipping jack" among the Negroes was an old Negro preacher, nicknamed "St. Albans." It was he who applied the "cat o' nine tails" for from 10 to 150 lashes according to the "kangaroo" judge's decision or the findings of a jury when an offender was found guilty. At one time we saw him whip another Negro until his shoes were full of blood.

"Why don't you stop them?" I asked the jailor.

"Can't afford to do it, for if I do they will become unmanageable," he answered.

"Bull!" I replied, "you mean to tell me that when such brutal treatment is accorded men under confinement that they are made better by it?"

"Don't know about that," he said, "but we reason that it is better to let them handle one another than to be constantly called upon to settle some internal dispute or infraction of the rules."

He would sit back and laugh while one of these whippings was in progress, and judging from his actions he enjoyed it.

Our continuous incarceration necessitated someone acting in our places, especially as president and secretary-treasurer. Keeney designated M. L. Haptonstall to act as president pro tem and I called upon Isaac Scott to act as secretary in my place.

Someone suggested, "Well, Keeney and Mooney are gone and it evolves upon us to decide who will fill each of these offices. They will never have the chance to occupy their positions again." Immediate dissension was the result because they could not agree upon who should have this office or who should assume the duties of that one until the news of their maneuvering reached us in the Kanawha County jail. After several of them were called in and told where to get off their zeal to divide offices slowly ebbed away.

Prisons will not retain the mind of the philosopher. When the body of a man possessing an active brain is imprisoned behind stone walls and bars of iron, his mind soars forth into the sunlight. He converses with his loved ones and comrades even though unlimited space separates him from them. Before the body of a human being can be enslaved, those seeking to enslave the body must first enslave the mind.

Eight out of every 10 human beings who are imprisoned for petty or other offenses are never able to look at the stars. When they are faced with the inevitable they hang their heads and their eyes become focused upon the mud. Because they are equipped with an inactive or dormant brain, they are never again able to lift their heads and gaze upon the stars.

Kanawha County jail was our domicile until Christmas Eve, 1921. During the afternoon of the 24th, Keeney called Sheriff Chafin of Logan County on the telephone and asked him, "If we give bond here tonight, will you hold up those warrants from Logan and permit us to spend Christmas with our families and meet you on C. & O. train No. 3 in Huntington on December 26th?"

"I sure will," Chafin replied. "Go home and be with your families through Christmas and meet me at Huntington the day after."

We gave bond about 11:00 p. m. and this gave us about 36 hours at home before our entry into the lion's den. The psychology of that act did much to secure our persons against violence when we were taken to Logan County.

We met Sheriff Chafin as agreed and with him were several deputies. They searched our persons and baggage in the wash room of the C. & O. depot at Huntington, and when we boarded the train for Logan we occupied a smoking room in one of the coaches, the deputies blocking the entrance. When we arrived at Logan, several additional deputies met us at the depot and acted as an escort until we were inside the jail.[1]

The jail was presided over by "Dad" White, often alluded to as Squire for he had previously served one or two terms as justice of the peace. I never learned his first name while there for I was not interested to that extent. As we entered the jail we were conducted along a narrow hallway to a row of cells in the back of the jail.

At the end of this hallway where we turned to the left lay a blood-stained mattress, and we were told that on that mattress was where the man died who was killed during the armed march early in September. They observed closely to ascertain if the sight of this gruesome reminder of how a human life had been snuffed out made any impression upon us as it was pointed out.

For my colleagues upon that occasion I cannot speak; I do not know what their reactions were. I can only speak for myself. Despite the fact that according to the situation in which I found myself I was classified as a criminal by the very society that had prepared my environment for me, and the further fact that society regarded me as one upon whom it had failed in its teachings concerning the way to live, I meant to assume the role of showing these gangsters and killers how to die. The man whose blood had stained the mattress was accused of trying to break jail by attacking the jailor's son.

Greetings from dozens of miners in each line of cells met us at every turn and we knew almost every one of them personally. One of them called out loudly, "Boys, they always said that if we went to jail they would go with us, and here they are."

Rev. J. J. Jeffries greeted us when we mounted the stairs to the second tier of cells. His broad, good-natured grin, his prayers and songs were responsible for the morale among many of the imprisoned miners.

We were placed in a row of cells along with Jeffries, two brothers by the name of Akers, Thomas Thompson, and another man who posed as a prisoner but in reality was a "plant" or provocateur. He gave his name as Thompson, but I never learned his first name.

As soon as we entered the cells Thomas Thompson, who was from Dry Branch on Cabin Creek, managed to warn each of us to be careful of what we said while in earshot of the "plant."

"No need, Tom," I replied, "I spotted that well groomed 'dick' immediately after we were locked in. If he ever soiled his hands in the coal mines he did it for some other cause than wages."

The "plant," as he was referred to in whispers, was a perfect specimen of physical development. The trained athlete was stamped all over him from his massive head of dark hair to his well-groomed feet. In every instance when he would engage in conversation with me, he would lead up to machine guns. My ignorance of these implements of industrial warfare was not just professed, it was real. But he would invariably lead up to that subject.

For three days following our confinement in the Logan County jail, we were placed on review for the benefit of the gunmen. Singly and in groups of two and three they were brought up and the jailor or some person in authority would "bawl" out, "Keeney, Mooney, and Blizzard to the front, these men want to look you over and see how many horns you have."

These bleary-eyed gun toters would scowl at us from outside the bars while the official would point out each of us and designate us by name and position. We met their scowls with a smile and greeted them with "Hello, fellows, how are you anyway?" An inaudible grunt was the only response from the majority of them, but occasionally one would come forward and clasp our hands.

Many of them would caress the handle of a protruding gun and lick their lips as though thirsty to shed human blood. While observing these gunmen, my mind wandered back to Maxim Gorky's creatures that once were men. But, no, I reflected, that would not apply. "Creatures that might have been men" would be more fitting.

One of these gunmen was unusually revolting or would have been to a timid person. He was six feet three or four inches in height. He weighed approximately 200 pounds and was dressed in the uniform of the state police. Two heavy Colts hung low from his belt and his face was distorted and misshapen with a hideous scar. During his scrutiny of the three of us he backed against the opposite wall and continuously fingered the handle of the Colt which hung from his belt on the right side. His malevolent grimaces in our direction increased his hideousness. When time arrived for their departure, he pulled himself loose from the wall with a sigh of resignation as though he had been waiting for a signal or sign to "slaughter them where they stand." He reminded me of a "pit bull" when two insurmountable barriers prevent him from sinking his fangs into a helpless pooch.

After the third day of this inquisitional inspection, I said to the jailor, "Dad, haven't you anything in this county but gunmen?"

Dad White looked at me with one of his sardonic grins, "Yes, Mooney, I guess we have at that."

"Well, if you have, I surely would like to see some of them for I never was so sick of looking at guns and gunmen in my life," I told him.

Keeney nudged me in the ribs and said, "Shut up, fool, you will get us all shot."

"Don't give a damn," I replied, "I'd just as soon be dead as to be in this shape anyway, so what's the difference?"

Next day Dad White permitted two railroad engineers and their wives to visit us and they brought a box of cigars, a carton of cigarettes, and a basket of fruit.

"See what a little mouthing did for us?" I chided Keeney.

Breakfast in the Logan County jail was a joke, lunch was a pun, and dinner was an anecdote. For breakfast we were given one tablespoon of rice, the same amount of apple butter, two slices of stale bread, and a cup of black coffee.

Luncheon consisted of about one-half teacup of navy beans (unseasoned), stale bread, and coffee. If any of the breakfast menu was left upon the tin plates, the beans were poured upon the rice and apple butter and thoroughly stirred.

Dinner (supper as we termed it) was composed of more beans and occasionally a piece of half-rotten meat. Rev. J. J. Jeffries often exhibited a piece of beef jaw that was served him with pieces of hay yet between the teeth. He also had a handful of broken glass that had been removed from the food served him.

The miners' defense attorneys contemplated using the beef jaw and broken glass as exhibits in testimony at the trials, but later decided not to do so.

John Stollings was scarcely out of his teens when the strike occurred. He was arrested and confined in the Logan jail during or immediately following the armed march.

Let him tell the gruesome story in his own way:

"After I was put in jail, they took me down in the basement and stood me against one wall. The gunmen lined up against the other wall, drew their guns, and asked me if I were ready to confess. I told them I didn't know anything to confess. They fired shot after shot close to either side of me, up and down the wall, the bullets splitting against and chipping the concrete from the wall and the lead and concrete slivers stung my back.

"'Confess, God damn you,' they said. I again told them I knew nothing. Then one of them caught me by the hair of the head and drew a butcher knife across my throat from one ear to the other just hard enough to draw blood.

"See that scar," and he leaned from right to left and a scar was visible from ear to ear. Between sobs he related his experience while the third degree was being applied. "I finally told them that I would tell them anything they wanted to know," he said.

By this method they forced a fake confession from John Stollings that was extremely nonsensical, and the boy was a complete physical and nervous wreck.

Through pity, because of his physical and mental condition, the defense did not use him as a witness, despite the fact that he repudiated the confession obtained under duress.

The prosecution did not use him as a witness as they were afraid of the cross-examination.

During the day of December 31, someone overheard one of the gunmen remark that "we are going to put the fear of God into them 'red necks' tonight." About midnight a fusillade of shots broke out around the jail, and from the direction of the courthouse the rat-tat-tat of machine gun was heard. Occasionally the din was pierced with what was meant to be a blood-curdling yell. But they did not take us by surprise; we were awake and listening.

After the bombardment had been carried on for several minutes, we began calling out to them, "Put up my board, you shot high to the right, low to the left, hey you are cockeyed, etc." These calls were in the vernacular of the old time Thanksgiving shooting match. Soon after we began teasing them they ceased wasting ammunition. One or two of them had the nerve to come through the jail and inquire if anyone was frightened.

On January 3, 1922, Keeney and I applied for bail. This application was made upon the assumption that if we could make a sufficient showing before Judge Robert Bland we might be admitted to bond.

During the examination, which was conducted by "Con" Chafin, prosecuting attorney for Logan County, and T. C. Townsend for the defense, Sheriff Don Chafin stood near drinking in every word and a court reporter jotted down the testimony.

Judge Bland denied the defense motion to admit Keeney and me to bond and we were re-committed to jail.

S. B. Avis was considered by the miners the most venomous and malignant prosecutor in the state at the time of the Matewan massacre in 1920. He had been retained by the Baldwin-Felts Detective Agency to prosecute the defenders of Matewan, and was slated to prosecute the miners and their officers in the trials growing out of the armed march. However, during the interim between the armed march and the trials, he and Attorney C. E. Goettman were enjoying a game of golf on the Edgewood course at Charleston. A thunder shower interrupted the game and the two players ran under a shed for shelter. A bolt of lightning struck the shed, killing Attorney Avis instantly and injuring Mr. Goettman.

C. W. Osenton, of Fayetteville, being next in reputation and formidability as a prosecutor, was employed by the Logan County authorities to prosecute the indictments growing out of the armed march.

Following the application of Keeney and myself for bond on January 3, 1922, we, along with scores of others including H. W.

Houston (chief counsel for the defense), were named in additional indictments which included treason. Despite the fact that I had subscribed for the Charleston *Daily Gazette* upon the promise of the jailor that if I did so I would receive every number, this was the only time we were allowed to see a newspaper while in Logan County jail.

The issue announcing the treason and other indictments was delivered to us. The charge of treason was the 14th indictment against Keeney, Blizzard, and me. The docket was getting topheavy, so they must have decided to quit. We reasoned, and were later informed, that the treason charges were Mr. Osenton's contribution to the pseudo prosecution with the destruction of the miners' union as the ultimate goal.

Early in January, Sheriff Chafin took Keeney to Huntington where a conference was held with Samuel B. Montgomery and others looking forward to an agreement on a change of venue to another county where the sentiment would not be so strong against the defendants.

Nothing developed from that conference with the exception of a suspicion that the prosecution was making an effort to inveigle us into a position where extrication would have been impossible.

Several days later our wives came to see Keeney, Blizzard, and me. Before they were admitted to the jail, I was cautioned not to permit my wife to deliver any written message to me. "Uh huh!" I said. I knew that if my wife had a message for me she would find some way to deliver it. When they entered the jail, she came to the bars and as she kissed me, she slipped a note into my hand.

"Careful," she cautioned.

I palmed the note, read it after she had gone out, then tore it into minute particles and threw it into the sewer. The note was from Defense Attorney C. J. VanFleet and contained the following advice:

"Refuse to sign any agreement or any other document without the advice of attorneys. Purported agreement for change of venue is only a trap." So I served notice on everybody concerned that I would refuse to sign anything unless the defense attorneys were present.

Among the inmates, a choir had been formed and was led by Edgar Holstein and others. Each night several hymns were sung and prayer was led by Jim Jeffries. He prayed as one who believed in the effects of prayer and sang as though he enjoyed it. Songs and prayer were the only diversions permitted. The mere possession of a deck of cards or a pair of dice meant confinement. When the elite of Logan gathered in church they thanked God that by the misapplication and distortion of manmade laws and because of the activities of hundreds of hired professional killers, operating in violation of all law, we were compelled to listen to their hypocritical ravings through a sound box connected between the church and jail. Lengthy prayers, spiced with thanksgiving, were offered to God for delivering the "red necks," "strikers," "marchers," and "bolsheviks" into their hands, and by the time the sermon was concluded, the "cooties" would decide that meal time had arrived; so "the battle for self-preservation" would begin and continue until sunrise of another day.

This monotonous procedure of fighting against resignation to fate, strengthening the strong to be stronger, continued until January 18, 1922.

"All get ready for a trip to the courthouse." This cry echoed through the corridors of torture and fear, and immediately all was bustle and confusion.

Early afternoon the call came for a general line-up, and from the jail to the courthouse we were marched through two lines of gunmen, standing from one to four in depth. Keeney and I led the procession followed by William Blizzard and Forest Payne. One hundred and three in groups of two passed into the courthouse.

Every three or four steps we would hear a growl from one of these minions of law (?) and look to right or left to observe a hand gripping a gun and an angry snarl escaping from lips that resembled animal more than human. When we were but three or four steps from the courthouse entrance one gunman said, "There's the god damn sons of bitches now, let's finish up the job right here."

Payne turned to them and smiled, saying, "All right, boys, burn your powder; that's what she was made for." However, the gunman was prevented from going into action by one of his colleagues.

Judge Robert Bland stipulated the amount of bond required as each name was called, setting Blizzard's bond at $15,000.00, Keeney's and mine at $10,000.00 each, and the remaining 100 applicants from $5,000.00 down to $1,000.00.

When the applicants walked over to the clerk to subscribe their names, Don Chafin increased many of the $1,000.00 bonds to $1,500.00, the $1,500.00 bonds to $2,500.00, and some of the $2,500.00 bonds to higher denominations.

This action was taken by Sheriff Chafin in face of the fact that Judge Bland had set the bonds in open court. Immediately following this procedure, we were conducted to the train that was to convey us to Huntington.

En route to Huntington, we were escorted by Sheriff Chafin and a score or more of his deputies. The second or third stop out of Logan, Sheriff Chafin got off the train and came back with a bottle of corn "likker" and asked me to have a drink. We walked back to the lavatory and when he pulled the cork, I hesitated. He laughed and said, "Fred, I'll take a drink first to show you it is all O. K." So the two of us killed most of the pint.

Our wives met us at Huntington, and Keeney, Blizzard, and I decided to remain there for the night and go on to Charleston the next morning. Thus ended the 109-day stretch in jail for Keeney and me. Blizzard had not been in quite so long for he was immune from the Mingo County incarceration.

## Chapter X

When we again assumed our duties as officers of District No. 17, a call was issued for a special convention to meet in the county courthouse in Charleston. During the interim between that call and the assembling of the convention, Frank W. Snyder, editor of the *West Virginia Federationist*, William Harris, president of the West Virginia Federation of Labor, and myself went to New York City and returned by way of Cleveland, Ohio.

Our mission was to enlist the aid of the big New York unions and the railroad brotherhoods in donating funds for defense of the many indicted coal miners. While in New York I purchased a detective's dictograph from the Dictograph Corporation of America.

We met the executive board of the United Garment Workers and secured a donation of $1,000.00. From New York we went to Cleveland and interviewed the presidents of the railroad brotherhoods.

Prior to that time the railroad engineers, of which Warren E. Stone was president, had entered the banking and coal mining businesses. They owned mines in Kentucky and had purchased coal property in the Coal River section, Boone County. The Boone County property was yet in the development stage; no coal had been shipped up to that time. When asked what their attitude was going to be toward signing an agreement with the United Mine Workers of America, Mr. Stone replied, "Just as soon as we are ready to ship coal, I want you fellows to organize my employees and we will sign a contract."

Imagine my surprise some time later when the subject of a working agreement was broached. They flatly refused to sign and declared for the open shop.[1]

Upon our return to Charleston, we began to prepare for the defense of the many miners who had been granted a change of venue, transferring their indictments from Logan to Jefferson County. The date of arraignment was set for April 22, 1922.

In the meantime the prosecution, under the guiding hands of Attorney C. W. Osenton of Fayetteville, A. M. Belcher of Charleston, and Robert Coates, a W. J. Burns detective of Pittsburgh, was preparing for one of the most relentless prosecutions in the history of industrial disputes.

I was assigned the duty of making an investigation of the private lives of each juror or prospective juror. By whom were they employed? And at what vocation? To whom did they owe money and what was the nature of their debts? What political favors had been granted to them and by whom? What was their religion and attitude toward organized labor, if obtainable.

Prior to the date set for trial, I went to Charles Town and conferred with James M. Mason, Jr., an attorney whom the defense had employed because of his ability and thorough acquaintance with the citizenry of Jefferson County.

Mr. Mason was a typical southerner and believed that the Constitution of the United States was in full force and effect. He was skeptical concerning the reports of how "Czar" Chafin ruled Logan County. He said to me, "Mooney, I am going into Logan County incognito and see for myself. For I cannot believe that there is any place in the U.S.A. that a man's person and property is not safe against unlawful search and seizure."

Mr. Mason later narrated his experience in Logan County:

"I went aboard the train at Huntington and selected what I considered an inconspicuous seat and proceeded to make myself comfortable for the trip to Logan. Some time after the train pulled out, several fellows passed through the coach and looked me over.

"Upon my arrival at Logan, some of the same group followed me to the hotel and there being no law against registering under an assumed name, I signed a name and was escorted to my room.

"So far, so good. After I had made my toilet, I proceeded to lock my door and go out for the evening repast. Before leaving my room I arranged my traveling bag until no one would be able to open it without my knowing about it.

"Imagine my surprise when I returned from the evening meal to find a man in my room, one in the hall, the door standing open, and my baggage rifled. I remonstrated with them and mildly inquired, 'What is the meaning of this, gentlemen? What are you doing in my room?'

" 'We want to know your business here,' they replied.

" 'Is a man compelled to state his business just because he enters this town and registers at a licensed hostelry?'

" 'Yes, in a way he is,' I was informed.

"But evidently both of them soon realized that they were making a mistake and slowly backed out and I was not molested again."

Seeing is believing, and Mr. Mason became a strong champion of the cause of the miners.

Jefferson County was typically southern. It was the birthplace of many historic characters. The people of that county received the defendants, witnesses, legal talent, and all others with that brand of hospitality attributable only to "Dixie."

William Blizzard was arraigned and tried in the same courthouse in which John Brown was tried and sentenced to be hanged. Both men were tried for treason.

The jury of 12 men, who sat and listened to the arguments relative to a bill of particulars, to the admissibility of this or that testimony, and to the evidence of approximately 200 witnesses, received the sympathy of the entire community.

This trial began on April 24, 1922, and continued until the evening of May 27, 1922.

Following our admission to bond on January 18, 1922, and after the return of Snyder, Harris, and me from New York, two dictographs were put to work. Later a third one was located by me and put to work without any of my co-defendants having any knowledge of its whereabouts or operation.

Through the use of these sensitive machines, we were able to learn every detail concerning how the prosecution witnesses were being trained in the subtle art of perjuring away the lives and liberty of men.

Each witness was given a typewritten statement to memorize. After he had studied it for some length of time, he was put on the witness stand, questioned, and cross-examined by different attorneys until he showed signs of being able to remember his part. He was put on the payroll, given a stipulated amount, and promised the balance due him when he had finished on the witness stand.

By this method we were able to circumvent or destroy much of the bought and paid for testimony.

During the armed march of the miners many circulars and pamphlets found their way to the marchers. We eventually learned that two communists had ensconced their hides (as they always do) safely behind the walls of the Washington Hotel in Charleston and were having the circulars printed and then sending them into the lines by "marchers" who did not realize their full significance.

While I was under cross-examination, Mr. A. M. Belcher, one of the leading attorneys for the prosecution, continuously harangued me concerning those circulars. During the cross-examining at Lewisburg[2] a circular, of which I was the author, was introduced in evidence. Handing me a sheet of paper, Mr. Belcher asked, "Did you write this circular?"

After examining the circular, I answered, "Yes, sir, every word of it."

Q. "Do you mean to sit there and tell this jury that you wrote every word of this document?"

A. "Yes, sir, every word of it."

Q. "Where did you get this language attacking the courts of this country? I'll read it to the jury: 'The greatest and most powerful enemies of organized labor in the United States today will be found sitting on the bench wearing the judicial ermine.'"

A. "That statement is an exact quotation from a speech made by you to the delegates of a United Mine Workers Convention in Indianapolis when you were on the payroll of the union as one of its legal talent."

To use slang, Mr. Belcher's "feathers fell." He quit cold. C. W. Osenton finished my cross-examination.

Publicity was not lacking. The newspapers of New York, Baltimore, Washington, Cincinnati, Charleston and many other cities were represented.

From the date of our entry into Charles Town, until the day of Blizzard's acquittal, the sentiment was overwhelmingly in favor of the defendants. The evening the jury rendered its verdict, the American Legion put on a "blow-out." They paraded through the streets carrying Blizzard on their shoulders. Before the "blow-out" was concluded it developed into a disturbance resembling an orgy and public sentiment changed overnight.

The trials of Walter Allen, J. E. Wilburn, John Wilburn, and another defendant followed that of Blizzard. All resulted in convictions. The defendant whose name I cannot recall was convicted and sentenced to serve 99 years in Moundsville Penitentiary. Shortly after his conviction he was taken out of jail about 2:00 a. m. one morning, transported to Harpers Ferry, placed on board a freight train and made his way to Athens, Ohio.

Walter Allen was released on bail pending a new trial and disappeared.

The Wilburns were sentenced to serve terms of 10 and 20 years in Moundsville but were later pardoned by Governor E. F. Morgan.

Following these convictions, charges and counter charges of bribery and jury tampering were hurled back and forth.

Changes of venue were sought and the cases against C. F. Keeney were transferred to Berkeley Springs. On the date set for trial, some of the state's principal witnesses "blew up" and the prosecution was not able to prepare a bill of particulars.

All other pending cases were transferred to Lewisburg in Greenbrier County.

Blizzard was again arraigned and tried for the murder of Deputy Sheriff John Gore of Logan County. This trial resulted in a hung jury. When the jury reported in this case it stood 10 for acquittal and two for conviction. Again, charges of jury tampering were hurled back and forth by both prosecution and defense.

One G. C. Hickey, miners' witness and sympathizer, was caught handing a note to one of the jurors. He was sentenced to serve a term in jail.

Changes of venue were again sought and the remaining cases were transferred to Fayetteville, seat of Fayette County. Every defendant involved was skeptical of Fayette County excepting myself.

I consistently argued, "We are now in a county where industrial struggle is thoroughly understood, and if there is one place in West Virginia where a fair and impartial trial can be had, it is here."

When C. F. Keeney was arraigned, I had made my usual check-up of the prospective jurors. The "strike" was ready when the defense attorneys arrived on the scene. Several days after the trial was under way, G. C. Hickey.

"What are you doing here?" I asked him.

"Blizzard sent me up here to see what can be done to help Frank out," he replied.

"The only thing you can do for Keeney is to injure his chances for acquittal," I told him. "Now you get into a room, stay in hiding until tomorrow morning, take the train back to Charleston, and stay there until this trial is over."

He took my advice and left for Charleston the next morning.

Duty compelled me to leave Fayetteville before the trial terminated. The joint conference between Coal Operators of Northern West Virginia and District No. 17 of United Mine Workers had been called to meet in Baltimore. As Keeney was on trial, it was my duty to assume leadership of that conference. It was in full swing when the news came that Keeney was acquitted.

Upon our arrival in Baltimore, the dissension that had been growing among the officials of the two northern West Virginia sub-districts was more pronounced than ever. Every sub-district official in Sub-districts Nos. 3 and 4 had turned into a calamity howler.

After the conference had gotten under way and following every adjournment at noon or at night, every sub-district official would say, "Fred, you had just as well adjourn this conference sine die and go home, for these operators are not going to sign an agreement. It's useless and hopeless."

Eventually this became monotonous to me and one evening while about everyone was present I turned on them. "What is the matter with you fellows, anyway?" I said. "Have you lost your nerve, turned yellow, or what? If you fellows whom the miners elected to represent them are incapable of doing what the men elected and are paying you for, why don't you be men enough to go home and tell them so? Admit to them that they made a serious mistake in selecting you and ask them to send someone here in your place that can at least rate a few degrees above an old mid-wife at a sewing bee.

"Now get this, every one of you, I am not going to adjourn this conference. If it is adjourned sine die, the operators will adjourn it, not me. If there is one or more of you who is so low on physical or mental stamina that you doubt your own ability to go through with it, let me have your resignation now and the executive board can act upon it later.

"I'll keep this conference in session until the 4th of July and start up again on the 5th, if these operators do not sign an agreement or walk off and leave us sitting here. If they walk out, the blame for failure will rest with them not me."

Mr. Percy Tetlow, later chairman of the coal commission, was present when this occurred.

The next evening a wire came to me that Keeney had been acquitted and that he would arrive in Baltimore the following day.

The joint conference was conducted to a successful conclusion and an agreement was signed.

A convention of delegates from Sub-districts Nos. 3 and 4 was called to meet in Fairmont to ratify or reject the work of the scale committee. Upon our arrival in Fairmont we met the same doubt and dissension that had been prevalent among the officials at Baltimore. "They are not going to ratify it," they told me.

A. D. Lavinder, one of the best friends I've ever had, said to me, "Old boy, I don't like to discourage you but these delegates are not going to ratify that agreement."

"Wait and see," I admonished.

I was the only man, of the entire official representation, who was firm in his belief that those delegates would ratify the agreement. Keeney, Tetlow, several of them spoke, but not a response. When Mr. Tetlow concluded his remarks and sat down he whispered to me, "Fred, it looks doubtful, doesn't it?"

"No," I replied, "when this situation is properly gone into these delegates will ratify."

I was the last one to take the rostrum in defense of the work of the scale committee. When I concluded there were only seven votes against ratification and each of those requested in turn that his vote be changed from "No" to "Yes," making ratification unanimous.

The information had been imparted to me, prior to this conference, that the operators were not going to fight the union. They would in all probability sign a joint contract and after it was signed let the mines remain idle or systematically close them down until the miners were starved into acceptance of whatever terms the operators offered.

This information proved to be correct and within a period of 90 days the miners were clamoring for the chance to work for bread at any price.

The trial and acquittal of C. F. Keeney concluded the private prosecutions. Ex-Supreme Court Judge Harold A. Ritz was employed by those who were financing the prosecutions to examine the remaining cases and either try or nol-pros them. All remaining cases were nol-prossed.

The defense fund, garnered by contributions from all parts of the western hemisphere, was exhausted by this time. Only sufficient funds remained to pay for the printing and mailing of 600 copies of a financial report of receipts and expenditures of the fund.

Copies of this report were mailed to international unions, state federations of labor, district councils, and central labor unions throughout the United States, Canada, Mexico, Hawaii, and the Philippine Islands. A circular was sent along with the report requesting each representative body to file the report for future reference.

As previously stated, the contributions to the Mingo County Defense League totaled $46,000.00. Twenty-six thousand dollars of this amount was expended up to the time William Blizzard was acquitted at Charles Town.

Legal defense funds have a peculiar psychological effect upon officers of a union. At one time the executive board of District

No. 17 passed a resolution demanding that I, as treasurer of the legal defense fund, turn all finances over to the executive board. I plainly told them "to go to hell."

"We will institute mandamus proceedings and take it," they threatened.

"Try it," I replied, "and I'll give bond and hold it despite your efforts. In addition to that I'll show why you want to get your hands on it."

That seemed to end the matter for the time.

\* \* \* \* \*

Following Keeney's acquittal at Fayetteville and the conclusion of joint wage conferences with those of the operators who would meet in conference and sign up, mines began to close everywhere. Suffering became intense among the miners.

They were bled white by persecution under the guise of prosecution and left to fight their own battle, as they had always been left to fight it, by the signing of what was known as the "Four Competitive States" agreement. Those involved where operators refused to sign agreements went on strike. Those where agreements had been signed were forced into idleness because the mines closed down.

Keeney and I discussed the possibility of working out a solution that would have as its chief aim the perpetuation of the union. In pursuance of this idea, some of the leading operators were approached and asked just how far they would go in an effort to maintain joint relationship.

We were reliably informed that if the $1.50 per day, which had been forced into the agreement by "wild cat" strikes, was relinquished the entire state could be signed up.

The big boss was consulted and this proposition put up to him. He tabooed it right now. "No backward step" was his slogan.

No backward step was taken, but the ground lost by the union became a landslide into the gutter for the miners. From a proud and democratic institution composed of 615,000 men, its membership dwindled to less than 100,000 dues-paying members.

In West Virginia, the miners were beaten down to the dregs and contracts were signed for tonnage rates far below the pre-war level. This condition prevailed until the birth of N. R. A. following the election of Franklin Roosevelt.

Following the refusal of John L. Lewis to permit the officers of District No. 17 of West Virginia to do anything other than fight what everyone knew was a losing battle, going further into debt every day, we called a conference of district and subdistrict officers to meet in Charleston and help formulate, if possible, a policy that would relieve the suffering among the miners.

About this time, a general rumor was set afloat that "if Keeney and Mooney would resign the operators would sign contracts for the entire district."

When the official conference met at headquarters it was soon evident that no three men in the district were together. We decided to lead the entire outfit up to the international executive board. A blanket resignation was signed and we were called before the international executive board at Indianapolis.

John L. Lewis was bitter in his denunciation of the miners' union officials in West Virginia. He berated us for trying to shoot the organization into the state and for the indebtedness incurred while trying to secure working agreements, despite the fact that the economic condition was a direct result of pursuance of policies forced upon the outlying territories by him.

The blanket resignation signed by the officials became effective July 15, 1924. Mr. Percy Tetlow was appointed provisional president and William Thompson became secretary-treasurer of District No. 17.[3] Weeks and months passed and the operators did not sign agreements. The year 1925 arrived and yet no sign-ups. The miners sank lower and lower, starving, striking, working, striking, starving until the advent of General Johnson and his Blue Eagle. But for the N. R. A. in 1933, the southern and southwestern counties in West Virginia would have continued under the iron heel of a system that had caused sporadic outbursts of violence and bloodshed for 30 years.[4]

In 1932, Ex-Mayor H. G. Kump of Elkins was elected governor of the state on the Democratic ticket. Inducted into office along with Governor Kump was Homer A. "Rocky" Holt of Fayetteville as attorney general. Soon after their induction into office, a situation arose which called for a ruling from the attorney general as to the legality or illegality of corporations paying private guards appointed by the sheriffs of the several counties involved. The ruling from the attorney general outlawed the use of private guards who were deputized by county sheriffs and paid by corporations.

Thus it evolved upon these two men to complete the work begun in 1918 by West Virginia's World War governor, John J. Cornwell, a Democrat. Governor Cornwell had been instrumental in eliminating the private guard system from a large area in northern West Virginia. However, the system operated undisturbed in southern counties of the state until the advent of another Democratic administration in 1933.

Let it be emphasized here that every Republican governor with which the State of West Virginia had been encumbered for 30 years had openly admitted that the private guard system was wrong and existed in violation of statutory law. Yet every one of them had winked at, ignored, or openly encouraged this system.

In 1920, 1922, and in 1924, I entered the race in Kanawha County as a candidate for the House of Delegates on the Republican ticket. I was nominated in all three races but beaten by a few votes in the general election. In 1924, after the Republican party had nominated Coolidge and Dawes, I jumped the Republican traces and supported Robert M. LaFollette for president. One of the crooked lords of the Republican party of Kanawha County called me upon the carpet for my nonconformance. He served notice upon me that if I expected to be elected to the legislature I would have to be regular, support the full ticket, and put up $1,000.00 as my contribution to the campaign. I plainly told him to "go to hell" and added that if I had to buy my way into the legislature and go there by reason of supporting men in whom I did not believe, I would stay out.

I walked out on them and made an independent fight. In 134 speeches, I told the voters of that county what the political lords had demanded of me and what I had refused to do. Packed audiences and overflowing fields greeted me in every nook and corner of the county. More votes were cast for me than for Calvin Coolidge, yet I was counted out by less than 100 votes. Against me were aligned both the Democrat and Republican political machines, the Ku Klux Klan, the Law and Order League, and the bankers and businessmen of the county. Yet despite these alignments, they were compelled to reach across into the Democratic ticket and pick up one of the Democratic nominees in order to defeat me. A man who stood high in the councils of the Law and Order League told me the next day after the election, "Mooney, we believe we have you beat. We spent $20,000.00 to keep you and William Blizzard out of this legislature."

"Well," I retorted, "it is at least some satisfaction to know that the bankers and businessmen and the two political parties will surrender that amount from their ill-gotten gains in order to prevent two men from being elected to an office as insignificant as the House of Delegates because they refused to be branded by the bunch of political crooks who control this county."

The Republican party in Kanawha County has never recovered from the effects of the 1924 campaign. The crookedness and rottenness with which it was infested stank to the high heavens and slowly but surely it was weeded out and was replaced by Democrats with the exception of the city government in Charleston.

I never on any occasion applied to the United Mine Workers for any consideration for a position since I presented my resignation in 1924. In 1925, I took the advice of Horace Greeley, who said, "Go West, young man, go West."

## Chapter XI

I had married again in November, 1917. However, this marriage did not prove successful and in July, 1925, we separated and obtained a divorce.

Taking my three boys and a nephew with me, I went to Albuquerque, New Mexico. I rented a house on University Heights and started working at the carpenter's trade. However, not much of this kind of work was to be found at that time in New Mexico. My nephew had secured a job in a logging camp in the Zuni Mountains, east of Gallup, and had urged me that should I make up my mind to leave there to be sure to come for him. About a month after our arrival in Albuquerque, I received a letter from a friend in Florida. He informed me that there was a boom in progress there so we packed our belongings and started to Florida.

We went to Clearwater and St. Petersburg. We encountered no difficulty in securing employment. While there, my wife wrote me that she had sued for a divorce, but if I would relent she would withdraw her suit and join me. I wrote her to charge me with any crime on the legal calendar and go ahead with her suit, as I would make no answer. I further told her that "Thanks to Florida, new associates, and new surroundings, I have regained my equilibrium and do not intend to have it disturbed by you or any other living woman." We stayed in Pinellas County for 16 months.

Leaving there we went west again to Abilene, Texas. Securing a job with the Sullivan Company of Ft. Worth, who were doing the excavation work for the foundation of a large hotel in Abilene, I shoveled mud for several days. I was down in a hole about 12 feet in depth shoveling mud up to a platform from which another mucker shoveled it up to the surface. One morning the foreman said to me, "Mooney, can you read a blueprint?"

"Fairly well, yes sir," I replied.

"Then you stay up here and assist me with this staking out," he answered. I helped him stake out the foundation and was not called upon to do any more mucking.

I had been making inquiries around town and learned that an engineering job was going to be open at the Hotel Grace. So I applied for and secured the job. But I wanted to go to Arizona where I had friends, and nothing short of that trip was going to satisfy me. Early in December, I quit the hotel engineering job and went to El Paso. I worked there about three weeks and pulled out for Phoenix just after Christmas. On January 2, 1927, we stopped in Camp Montezuma on East VanBuren Street in Phoenix.

In my many conversations with Mother Jones, she had often mentioned Governor George W. P. Hunt of Arizona.[1] The day after we arrived in Phoenix I drove out to the state capitol with the intention of seeing and becoming acquainted with Governor Hunt. I had with me a letter of introduction given me by Mother Jones addressed to T. V. Powderly of the United States Labor Department in December, 1920, when I applied for a passport to enter Mexico.

When I went into the governor's office he was dictating a letter. When the secretary handed him my letter and he had read it, he said to her, "You are excused until I call you." Turning to me he said, "Come in, young man, and sit down. Do you know that old lady?"

"Yes, sir," I replied, "since I was about eight years of age."

He motioned me to get up and putting his arm across my shoulders, he guided me across the office to a lace covered frame hanging against the wall. He parted the lace curtain in the center and exposed a large autographed photo of Mother Jones. Turning to me, he said, "You know, boy, I think she is unquestionably the greatest woman this nation has ever produced." Then he related to me circumstances of her advent into Arizona during a strike among the copper miners.

"What do you have in mind? Are you going to stay in Arizona?" the Governor inquired.

"It is too soon to answer that question, Governor," I said, "Give me a little time to look it over."

"Well, if you decide to stick around, look in on me once in a while, and if there is anything I can do for you, call on me," he said.

I thanked him and started looking the city over for a job. I worked three weeks on a small job in Phoenix. Having concluded the job and finding work scarce, we started for Yuma.

Before leaving El Paso, I had talked with the superintendent of the McKee Construction Company and he told me that if I would follow him to Kofa, Arizona, he would employ me. Kofa was located on a new line of the Southern Pacific Railroad, running along the southern border of the Harqua Hala Desert and north of the Gila River. We stayed overnight in Agua Caliente and started for Kofa. We soon learned that the new railroad had interfered with the desert trail in many places, cutting across the trail and leaving nothing but recently moved desert silt where the highway had previously been. After being compelled to dig our car out of the silt on several occasions, we arrived in Kofa, only to have the man I had conversed with in El Paso deny ever having seen me.

Working our way back to the main desert trail, we started for Wellton. Arriving in Wellton, we stopped in Fred's Tourist Camp for the night. Someone told me the next morning that there was a theatre going up in Yuma; we went to Yuma, but found the job was still in the prospective state.

We did not know just where to turn next, so we pounced upon a field of cotton belonging to a Mr. White. This was my introduction to cotton picking. About two weeks later the Seaboard Bond & Mortgage Company of Long Beach, California, started an enterprise five miles east of Wellton. I secured a job from this firm and worked there for about six weeks. Fourteen thousand pecan trees were planted, and 640 acres of land were seeded in alfalfa.

During this period we occupied a ranch house belonging to Jim Chappel, then sheriff of Yuma County. The ranch house stood on a knoll about one fourth mile southeast of the pecan farm. Evenings and Sundays were spent prowling into the dry washes and canyons north of the Gila River. We dug and panned in the pot holes along the washes for colorings of gold. Puma tracks were in evidence along the north side of the river as were signs of antelope coming at night for water. A wild burro was occasionally seen and once we found where a puma had captured a wild burro.

About five weeks after we located in the ranch house I was awakened about 4:00 a. m. by an ominous rumbling. I secured a flashlight and went out on the back porch. I saw what had, become a raging torrent in a few hours. Only a few scattered showers had fallen in that area within a period of three months. The snow had been melted in the Rocky Mountains with a spring rain. We were completely marooned (unless we resorted to swimming out). The river was a rumbling, tumbling mass of water, silt, mud, and desert rubbish. From the mesa on the south to the north foothills shore it was more than one half mile wide. Species of wild life were crowding onto high places. From the appearance of an ordinary dry wash the stream had assumed the proportions of a large river in a few hours. The water receded as fast as it had risen and by the second day the river was almost normal again.

On April 1, we left Wellton and came back into Phoenix. Contacting a man with whom I had become acquainted on my previous trip, I secured a job and rented a cabin in Camp Navajo at 16th and East VanBuren streets. I worked in Phoenix until Thursday, May 12. Someone had told me about the construction of a hotel in Prescott. We started for Prescott just before noon and camped in the pines south of the city for the night. The following morning I parked the car at the curb in front of the site on which the Hassayampa Hotel was being erected. When I stopped the car a man came walking down a plank which was used as a runway for wheelbarrows.

"Where may I find the superintendent?" I asked.

"I am he," he answered.

"Can you use another mechanic?"

"Yes, come out Tuesday morning," was the reply.

I thanked him and drove west to the Pine City Tourist Camp and stopped at the office. While I was talking to the owner about renting a cabin, he asked me if I would consider building a concrete bath and wash house to service the tourist camp. I told him, "Sure, if I can build it evenings and on Sundays."

"O. K. with me," he agreed.

Prescott was the cowboy capital of the world and was one place which I had wanted to visit for years. I steadfastly resolved to try to find sufficient work here to keep me busy until after the Rodeo on July 1st to 5th.

Late in July, I received a letter from my Mother saying Dad was ill, so I decided to return home to Charleston.

We arrived in Charleston on August 11, and I negotiated for a restaurant near the C. & O. depot. My boys and I operated the restaurant until April, 1928. One day during late March, my Mother came to town and stopped in to see me. I prepared her some food and while she was eating she observed me looking down at the toes of my shoes.

"Well, stormy petrel, when are you going again?" she asked.

"It won't be long now," I replied.

Gazing intently at me she mused, "Yes, when I see you looking at your feet I know you are thinking of going again."

These fathers and mothers. What shall we do when they are gone? I know and fully realize that it is inevitable that some day my parents must depart and yet I can't seem to make up my mind that perhaps they will precede me into the unknown. When you take your troubles to an acquaintance, he detests you and thinks of you as a squealer. A stranger ignores you; a lawyer gives you legal advice and collects a fee. A doctor prescribes for you, gives you pills, and exacts his pound of flesh. A psychologist will practice upon you the "science of the mind"; a preacher or priest will tell you to pray and trust in God. Good advice, but none of it will fill a hungry belly, clothe a naked back, or bring back a loved one. The magician will gaze into a globe of glass and while mumbling incoherent chants will paint florid tales of the good fortune that will befall you when time rolls around to a certain sign of the zodiac. The professor will recommend a course in public instruction to improve your chances in competition. The politician whom you supported and voted for will hand you a line of political gab, couched in carefully obscured linguistic terms, and refer you to his next understudy. Perhaps he will call the understudy by phone, send him a message by letter or telegraph, and tell him to handle you in his own way but to refrain from any commitments.

Not so with Mother and Dad. They will split the last penny; divide the last crust of bread; come to you in the dog-watch hours of the night; mortgage the old home place to pay a lawyer, if you are in trouble, the doctor, if you are sick. Have you committed a crime? A jury may convict you, but Mother and Dad, never. Have you become an outcast? Perhaps to the rest of the world but not at home. Have you hit all the back doors along all the routes you have travelled? Have you been called hobo, bum, beggar, vagrant, given a cold stare and had doors slammed in your face? If so, then go home for there you will find the latch string hanging on the outside of the door.

\* \* \* \* \*

My second eldest boy and I went west again in April, 1928. On our way we stopped over in Columbia, Missouri, and I worked several weeks there for the Shultz Construction Company, which was at that time erecting the Missouri Theatre. Leaving there when the job was just about completed, we trekked to Santa Fe, New Mexico. We arrived late in the night and put up at a tourist camp. I hurried through my breakfast the next morning because I wanted to go to the Pueblo Indian reservation to see the foreman concerning a job.

When I got up from the table an Indian vendor of bows and arrows who had been standing close by came over and sat down on the camp stool which I had vacated. "Ugh!" he grunted, and filling his plate he began to eat.

My boy gave him a quizzical look and said, "Big Chief, you've got your nerve, haven't you?"

"Ugh!" replied the Indian and kept on eating.

"Let him eat," I told the boy and left laughing at his discomfiture.

This was about the 22nd day of June and when I applied to Mr. Henry Peabody, foreman for the United States Government on the job, he told me, "Yes, I can use you, but not until after the end of the fiscal year June 30."

A dormitory and hospital were being constructed for the benefit of the Pueblo Indians. I was told to report for work on July 2. When I returned to camp the boy told me that the Indian had by grunting and use of sign language induced him to fry two extra eggs for his repast. We moved into the Sierra Vista Park that day and pitched our tent for the waiting period until July 2.

We stayed in Santa Fe until late in August. The job I was working on was nearing completion, so we began to make inquiries concerning new territory.

We left for Carlsbad and stopped over the first night in Murphy. The following day we arrived in Carlsbad and visited the caverns. I soon secured a job from a Mr. Toffelmire who was erecting a residence for Mr. Hendricks near the courthouse square and another for the city marshal, William Woods, only a few blocks away. While working in Carlsbad, we attended the cotton carnival at Roswell which is always a gala affair in the Southwest. On the day we visited the carnival we listened to a speech by Charles Curtis of Kansas, who was then running for Vice President on the Republican ticket in the Hoover presidential campaign. When the jobs in Carlsbad had dwindled down until the Hotel Caverna, which was being erected by the McKee Company of El Paso, was about the only building under construction, we left and went to Douglas, Arizona.

When we arrived in Douglas the Gadsden Hotel was in the process of construction by the Ware Ramey Company of El Paso. I went to work on this hotel as a carpenter and my boy secured a job in the Copper Queen smelter as a furnace stoker. It was about time for the hunting season to open in Arizona. After making some inquiries, I secured a hunting license so I would be ready when the time came. Each week-end was spent in some spot where game was plentiful.

On December 1, 1928, we packed up and continued northwestward into Phoenix. When we arrived there one of the largest structures ever undertaken up to that time was being erected near Phoenix. The Arizona Biltmore was rearing its magnificent head under the southern lee of Squaw Mountain. Into the net of high finance necessary to construct an edifice of this magnitude were drawn many substantial citizens of Arizona who woke up shortly after the dedication ceremony to find themselves numbered among the shorn lambs.

I worked on the building for several months and when it was finished a celebration was held by the promoters in a downtown hotel. The Arizonians who had assisted in the financing of this proposition were wined, dined, and toasted as the pioneers of progress in Arizona. Several days after this banquet was held, I visited with Governor Hunt at his home and during my stay our conversation drifted to the Arizona Biltmore.

"Pioneers of Arizona, eh!" the governor said to me. "The real pioneers of Arizona crossed the mountain astride burros with prospector's tools and camping equipment on the backs of other burros. What benefit do the people of Arizona expect to derive from a proposition of that kind? No, Mooney, I'll tell you what is going to happen to the people who put their money into that building. They are going to wake up one of these mornings with a bitter taste in their mouths that water will fail to wash out.

"The bouquets cast in their direction," he continued, "will turn out to be preliminary lilies to their financial funerals. If you wish to keep posted, watch the bankruptcy proceedings in Arizona for the next few months."

Following his advice, I watched the court proceedings and business failures bear out the governor's prediction. Governor Hunt loaned me several books, including *Oil* by Upton Sinclair.

Soon after the Biltmore job was completed I entered the San Carlos Hotel as engineer under Mr. Knotts as manager. This was my first introduction to air conditioning and I found it very pleasant work during the summer of 1929 when the temperature outdoors sometimes soared above 110 degrees. I had plenty of work to do but I always enjoyed working for Mr. Knotts. He was always in a pleasant mood regardless of what went wrong and he never forgot to be a gentleman.

Late in the fall of 1929, I left the San Carlos and went to work for the contractor who was erecting the Montgomery Ward building in Phoenix. After finishing there we went to Ajo, Arizona, and I started to work for the Calumet Arizona Copper Company.

Ajo is in a class by itself. While there we paid one cent per gallon for all water used. The water supply is obtained from a well 1200 feet in depth. The water from this well tests about 110 degrees Fahrenheit when it is pumped to the surface.

Finding it difficult to secure a place to live, we left Ajo and went to Douglas. However, work was scarce there and it was three weeks before I secured a job. One day a cow puncher with whom I had become acquainted on my previous stay in Douglas, said to me, "Mooney, why don't you go with Reagan and me after stumps?"

"Stumps? What do you mean stumps?" I asked.

"These walnut stumps he is taking out of the land through this country," he replied.

I went to see Mr. Reagan and arranged to go with him on his next trip. The cowboy above mentioned carried several aliases such as "Cowboy Jack," "Happy Jack," etc. His real name was J. E. Harris and I shall have more to say of him later.

When everything was in readiness we started for the San Pedro River and began searching for and securing walnut burl in the vicinity of Hereford and Fairbank near Fort Huachuca.

In the southwestern part of the United States and in Sonora, Mexico, about 80 to 90 percent of the black walnut tree grows under ground; in other words, the walnut tree is 80 to 90 percent stump and root. This burl was being collected at that time for the Purcell Walnut Company of Kansas City. We received $.18 per pound for it at that time. I was told that this company shipped the material to Germany where it was placed in giant lathes and turned into sheets of veneering. While this burl material was collected, it was handled with utmost care for if the bark covering was damaged, except where a root or limb was removed, it was culled at the point where it was taken up. If the damaged wood was accepted at all, it was at a reduced price.

We removed specimens of this burl from walls of canyons and bottoms of deep dry washes under circumstances which would seem impossible unless one understood the equipment used. A crane similar to that used by automobile wreck trucks was installed on a two-ton Chevrolet truck and equipped with about 300 feet of steel cable to which was attached what timbermen classify as a lizard.

In searching for this material one hunted walnut shrubs growing out of the ground. They ranged in size from mere sprouts to eight to 10 inches in diameter and reached heights 15 or 20 feet above ground. It was only on rare occasions that any portion of the burl appeared above ground. When a group of shrubs was discovered, shovels and mattocks were employed to remove the rocks, sand, and dirt which covered the hidden burl. As the uncovering proceeded the value of the burl could be determined. We removed specimens of this material along the canyons and dry washes which flowed into the San Pedro River which ranged in weight from 300 and 400 pounds to 4,200 pounds each. Many were beautiful in design having the appearance of toadstools or banks of flowing lava as though one layer had grown over another until the entire mass was curly.

Camping equipment was carried as an important part of each trip and water supply was transported in barrels and canteens. However, the tents were never set up unless the weather threatened rain or it actually rained.

When we returned from about a three-week trip, I secured a job erecting a store building for a barber by the name of Nelson. Upon completion of this job I entered the property of the Copper Queen smelter and went to work on a large reduction plant which was being erected there by the Phelps Dodge Corporation. At that time, Douglas was enjoying somewhat of a boom and carpenters and mechanics had been brought in from Los Angeles, San Francisco, El Paso, and various other places.

During my employment by the Phelps Dodge Corporation, a big conference of industrialists was called by President Herbert Hoover. Following this conference an announcement was made from Washington that all the big business corporation heads had agreed with Hoover to not reduce wages or lengthen hours during his term as president. The evening of the day this item came out in the press all the boys came from work in happy spirits.

"Did you see the good news?" they asked me.

"Yes, if you mean the newspaper report that there is to be no reduction in wages," I replied.

After they were all seated on benches for the evening rest usually indulged in before washing up for the evening meal, I looked at them quizzically and warned, "Boys, I dislike to throw a damper over your enthusiasm, but when President Hoover and a group of that kind get together and put out a report of that nature, you had better keep a weather eye out for the facts when the smoke screen clears away."

"You are crazy!" one of the group cut in.

"O.K.," I replied, "but wait a few weeks and see."

Lo and behold! Our next pay envelope revealed a reduction in wages from $6.01 down to $5.77. Soon after this occurrence the big reduction job was finished and to this day has never been used. A merger took place between the Phelps Dodge Corporation and the Calumet Arizona Copper companies and the Copper Queen was closed.

Following the completion of this job my boy and I erected a tourist camp for the Mr. Nelson for whom I had previously built a store house. When we had finished we returned to West Virginia.

## Chapter XII

We set up our tents on Davis Creek near Charleston. While we were encamped there Frank Keeney came for me to help him with the Reorganized United Mine Workers of America.[1]

Under the leadership of Alexander Howatt, John H. Walker, Adolph Germer, and others, a group of miners in Illinois had broken away from the Lewis union and made agreements with certain coal companies in Illinois.[2] Keeney asked me to join with him in the re-organizing work in West Virginia. My previous experience had enabled me to know personally several thousand coal miners in West Virginia. I was to work through the northern part of the state while Keeney took care of the southern end.

The United Mine Workers, as I had known them, were a social lot and in each community each individual knew everyone else. For instance, from 1917 to 1925 one could go to Cabin Creek Junction and ask anyone who happened to be present, "Where does George Carr live now?"

"Why he is still at Kayford," would be the answer.

This knowledge of one's neighbors and their whereabouts over hundreds of square miles was prevalent in all communities where the organization had spread in the years from 1917 to 1925.

When I came into Clarksburg, Fairmont, and the surrounding coal fields, I found the old timers scattered to the four corners of the earth. Some of them had entered other pursuits. Others had left the state. The mental attitude of those who could be located was depressing to say the least. They had drifted into a state of lethargy or hangdog attitude from which it seemed impossible to arouse them. The union had been almost completely destroyed. Only small groups, living in barracks in a few out of the way places on private property, were active in any capacity whatsoever.

A short time after entering the field, I billed and advertised a meeting on the old college grounds at Flemington. The meeting was going along nicely when in came several car loads of people led by Frank Miley (a Lewis appointee) and a Negro by the name of Nelson Beatty. They announced their arrival with hooting at the top of their voices and tooting auto horns. The tooting and hooting continued until they had pushed the speakers from the platform. After this had been accomplished, first Beatty and then Miley delivered what they considered to be speeches.

Their remarks were punctuated with vituperation and innuendo. At the end of each tirading sentence a line of praise for either John L. Lewis or Van A. Bittner[3] was injected. As I listened to the tirades being delivered by these men I began to wonder what had happened to the coal miners in northern West Virginia. Every word these men uttered from the rostrum relative to the facts concerning the previous administration of the United Mine Workers was a prevarication and every sentence loaded with charge and countercharge.

After listening to them for some time the thought occurred to me that these men were human parrots, that the language they were using had been placed in their minds, and that they were being carefully coached in what to say and when and where to say it. After I gave the attitude of these men some careful study and consideration, it dawned upon me that they were parroting Van A. Bittner. But for the inflections in the voice of each speaker being different, a listener would have believed that he was hearing one of the tirades called speeches which only Van Bittner was capable of delivering.

A member of the state police and the town marshal of Flemington arrived in time to prevent a riot from occurring that day. At that time a Mr. Jones, a coal operator from McDowell County, had announced his candidacy for the United States Senate. The speaker representing the Lewis faction charged that Keeney and I were in the Fairmont field to support the candidacy of Mr. Jones. Keeney and I had never even met Jones and knew nothing of him, nor had we contacted any of his field marshals or campaign managers. This charge sounded as though it had come from the campaign headquarters of M. M. Neely, who was a candidate at that time on the Democrat ticket for the United States Senate.

Following the episode of the Flemington meeting, I began to dig into the why and wherefore of this campaign of slime and vituperation and I learned that they were uneasy for fear I would stumble upon some of the facts relative to tipples and railroads being dynamited and tipples being burned during the previous strike. One day during a chat with a prominent lawyer in Fairmont, he said to me, "Mooney, if you could uncover the facts relative to some of the tipples that were bombed or burned and the railroad bridges that were bombed, you might build up some sentiment for your cause."

"Are you really interested in knowing who destroyed that property?" I asked him. "If you are, I believe the evidence placing the guilt where it properly belongs can be secured."

"You place the evidence in my hands and furnish me assurance that the men will go into court, stand up and testify and I will use it," he answered.

I had accidentally stumbled upon fragments of this evidence and had learned that at least one man and perhaps two had gone to the penitentiary for crimes they had not committed. I began quietly to ferret out the men who were in position to know the facts about these occurrences. When the subject was broached to them they would shake like an aspen leaf. The only condition under which they would agree to go into court and testify to the truth regarding these crimes was that they be given full protection and at the conclusion of the trial be furnished sufficient funds to enable them to migrate to a far and distant part of the United States or Canada.

Imagine my surprise when I again approached the attorney relative to this question and he did not even give me time to relate what I had learned but veered away from the subject altogether. At this point I dropped the inquiry and had nothing more to do with it.

I changed my tactics in the organization campaign and began to meet men individually, in pairs, and trios. For some time I did not undertake to hold another public meeting. It was hard for me to understand the tactics being used by the United Mine Workers' officials for I had been a member of that organization almost continuously for 19 years. I had always understood it to be founded upon democratic principles. I had heard its leaders boast of its high ideals, prate of its tolerance, and express pride in the fact that it was the only institution of any kind on the American continent that had solved the melting pot. I had never known it to be guilty of suppressing free speech or breaking up public gatherings, no matter by whom or for what purpose they were being held. I had always regarded it as a shrine into which men entered and were then freed from the restrictions so often practiced by intolerant public officials, thug-ridden coal communities, and communities governed by fanatics.

I quietly established several local unions in the Fairmont-Clarksburg-Grafton section and held numerous conferences in the backs of pool rooms, in private dwellings, and in obscure and out-of-the-way places. In many instances I came in contact with men who wanted to join but were afraid, not of the coal operators, but of the network of spies maintained by the United Mine Workers in each community.

One evening, along with Tony Costelac of Shinnston and Frank McCartney of Clarksburg, I went into Wendel. This mine had been solidly unionized in 1917 when a contract was secured with the Maryland Coal Company and had remained so until I left the mine workers in 1924. Between 1924 and 1930, the union had broken at Wendel as it had in other places and the men were working non-union. We found a couple of the old timers in houses on Scotch Hill and while talking matters over with them I said, "Can you get five or six of the others together so we can talk to them?"

"No," one man answered, "I don't know any of them."

"How about your neighbors," I inquired, "can you not call in two or three of these fellows who live next door?"

He hung his head for a few seconds, then looking at me said, "No, I can't for I don't know any of them."

As we started away I said to Costelac and McCartney, "This is poor material to work on. These men are either afraid to talk to one another or they have been beaten down until they do not care." This attitude was prevalent in each community we visited and the poverty, indigence, and squalor in each mining camp was appalling. The miners were working for wages less than the pre-war level and as for conditions, there were none except those dictated by the employer.

Several groups near Fairmont kept asking me to hold an open meeting in the courthouse and finally I consented. Early in January, I went to the prosecuting attorney and secured the use of the county courtroom in Marion County. I put out handbills and advertised the meeting for 7:00 p. m. on a Saturday evening. Prior to this time Van A. Bittner had delivered one of his usual harangues over WMMN, the radio broadcasting station in Fairmont. I tried to secure time at their regular rates to answer him over the same station. When I submitted a copy of what I intended to say, which was absolutely clear of the brand of slime, vituperation, and innuendo used by Van A. Bittner against Keeney and me and contained only a statement of facts, I was told by the management of the broadcasting station that I would not be permitted to deliver those remarks over their broadcasting facilities at any price.

I wired Adolph Germer to be present if possible and deliver an address at the meeting billed for the courthouse in Fairmont. When we arrived at the courthouse near the time set for our meeting, the courthouse was already filled and an attempt was being made to block all the entrances. However, I knew a back entrance and through this we entered and made our way to the rostrum. Several bus loads had been transported from Scott's Run, Wheeling, and Pittsburgh, at the expense of the United Mine Workers, to swell the crowd and capture the meeting. While our meeting was billed for 7:00 p. m., they had entered the courthouse at 6:00 p. m. and had full possession when we arrived.

I went to the sheriff and appealed to him, calling his attention to the fact that the prosecuting attorney had granted me the use of the courtroom on this particular occasion. His reply to me was, "If you fellows and Van Bittner don't quit your quarreling around here I am going to lock you all up." The prosecuting attorney was out of town and I could not locate him, so after again appearing on the rostrum and protesting against the tactics being used, we retired from the meeting.

After the meeting had been completely captured and several of Bittner's lieutenants had delivered their tirades of lies and slime, a committee was sent to the Fairmont Hotel to invite him to appear and address the gathering. Bittner could not command an audience at that time unless it was imported from distant points at what some of them told me was $2.00 per head and a bus ride. The imported audience would sit and listen and deliver their usual $2.00 worth of applause.

The coal operators never interfered with me nor disturbed me in any way. There was not even one instance where an operator asked me to leave his property or refrain from talking to his employees. This included the properties of the Consolidated Coal Company, the Maryland Coal Company, the Reppert Coal Company, Dawson Coal Mining Company, and many others. One of the operators was frank enough to tell me in a good-humored way, "We don't have to bother about you; the United Mine Workers will take care of the situation."

Late in January, a few weeks after the Fairmont episode, I decided to undertake another public meeting in Clarksburg. It was some time before I could get Walker and Germer to agree for me to hold this meeting, but they eventually gave their consent. As a prelude to this meeting, I rented the Carmichael Auditorium and received a receipt for the use of it from 3:00 to 5:00 p. m. on Sunday. I issued public statements challenging Van A. Bittner or any of his understudies to debate the issue from the public rostrum on that day. I went to the prosecuting attorney of Harrison County, to the sheriff, and to the state police at Haywood Junction, and told them what my intentions were. I demanded protection for the people who wanted to attend and hear Germer, Keeney, and me.

Mr. Carmichael gave me the key to the auditorium and told me to open it at my convenience. I waited until the appointed time, then turned the crowd into the hall, mounted the rostrum, and took charge of the chair. Sheriff Grimm and his deputies, Mr. Brown with a group of state police, and a group of city police were on hand to see that the hall was used by those who had paid for it and that the meeting was orderly for those who wished to attend. I had chartered streetcar and bus transportation to bring the people who were in sympathy with our efforts to the meeting on that day.

About the time the meeting had convened, several bus loads of people were brought in by the United Mine Workers, led by Frank Miley, Tony Teti, and others of the Lewis appointees. When the imported crowd had entered the hall, one of the mine workers' "hangers-on" offered a motion that Frank Miley, appointed president of District No. 31, act as chairman of this meeting. I ruled the motion out of order and using my pocket knife for a gavel announced to the gathering that the hall belonged to me for the afternoon. I told them it had been bought and paid for by the Reorganized United Mine Workers and that I intended to act as chairman, introducing the speakers, and those who wished to sit and listen, conducting themselves orderly, were at liberty to do so. To those who were there for other purposes than to sit and listen, I suggested that now was the time to retire.

After I had made this announcement, Sheriff Grimm rose from his seat and said, "Now gentlemen, this hall belongs to Mr. Mooney for this afternoon. He has asked for protection for the group who wishes to sit here and listen. I am a believer in free speech; no matter what a man has to say, I believe he has a right to say it if someone wishes to hear him. If any of you have anything else in mind than to sit and listen, you had better leave the hall now. We have a hoosegow down here and if you start anything there is enough of us here to take you and we will surely put you in it."

At this juncture, Frank Miley and his fellow appointees waved to the imported crowd to follow them and retired to the street. The meeting proceeded in an orderly manner. After the meeting was concluded we learned that Van A. Bittner had been safely ensconced in the Waldo Hotel waiting for the usual committee to call and escort him to the Carmichael Auditorium for the delivery of his customary declamation.

Not long after the Clarksburg meeting a judge somewhere in Illinois issued an injunction which placed the reorganized mine workers at a standstill for some time. Early in 1931, the reorganized mine workers consolidated with the United Mine Workers.[4]

## Chapter XIII

After attending the mine workers convention in St. Louis about the middle of April, 1931, I went to El Paso, Texas. I had re-married on January 3, 1931, after having been divorced for five years. My third wife was Miss Virginia Rowan of Fairmont. She and my youngest son, Harold, accompanied me on my journey west. We arrived in El Paso on April 23, and I began to look around for work. A few odd jobs were obtained, but none was to be had of any consequence. El Paso had not yet gone to the bottom of the Hoover administration ditch. It was on its way but it died more slowly than many other communities.

We stuck in El Paso until it eventually became industrially as dead as a mackerel. By July, 1931, no work was to be had except an odd job of a few hours or days at a time. On July 29, 1931, I was sitting in a park in El Paso reading the daily paper and was especially scanning the "Help Wanted" section. A man sat down near me and after making casual remarks concerning the weather, he asked me to let him see the paper I was reading. "Oh! I see, you were looking over the help wanted ads," he said.

"Yes," I replied, "I am a carpenter by trade, and I thought an odd job might appear in that column."

"Carpenter, eh? I have a small porch job up on the hill I would like to have done," he said.

"Be glad to do it for you," I offered.

"Let's get in my car and go up and look at it," he said.

Upon entering the car he remarked, "Well, let's run over and get a beer before we go up on the hill." We crossed the bridge into Juárez, Mexico, entered a saloon, and when we were seated he ordered two big ones. When the steins were emptied he called to the barkeeper to fill them up again. After drinking the second glass of beer, I started for the lavatory, but I do not remember ever reaching it.

For a period of several days everything was a perfect blank. When I returned to consciousness I was awakened by a chugging and swishing. Looking around in the dark I found what later proved to be a porthole and through it the stars were visible. It took me several moments to reason out that I was aboard a river or sea craft of some kind. "How did this come about?" I asked myself. "There are no craft of this size on the Rio Grande River."

Getting to my feet I began to feel my way around the walls and eventually staggered into steps leading upward. I was so weak that the effort of finding the stairway had almost completely exhausted my strength. I sat down on the floor and rested for several minutes to regain strength and relieve the throbbing in my ears. When about ready to continue my exploration, my plans were knocked "haywire" by the sound of voices from above. Two men came to the top of what later proved to be a companionway and after conversing in low tones for some time one of them started down.

I quietly made my way back where I thought my bunk was and while feigning slumber I looked and listened. He came to the bunk, leaned over me and by the light of a match peered into my face. "Still dead to the world," he muttered to himself. A few minutes later his pal came down the stairs and thinking I was still out they stood at the bottom of the stairway and discussed me at length. They had been paid $50.00 at San Diego to take me to Alaska or Kodiak Island and leave me stranded.

Plans to escape began to form in my mind, so on their next trip down I made it appear that their entrance had awakened me. I was permitted to relieve the demands of nature and then was given some food. The coffee had an unusually bitter taste. Soon after drinking it I felt a creeping drowsiness coming over me. When I again awoke it was late evening and I soon realized that I was on board some kind of a tramp freighter. Hunting through the compartment I found some old clothing and a pair of old shoes. Among the rubbish was a collection of wooden pins. Securing one of these, I crept up the stairway and looked about.

Everyone was busy except one man who acted as if he were a lookout. I watched his maneuvers for a short time and finally decided to make a try at a getaway. It was now getting dark as the sun was about to disappear into the Pacific. On the side of the boat where the loading work was taking place lights were being lit up. The shore line was near and on that shore was the fringe of a forest. Those trees looked good to me. While taking my bearings, I reasoned that if I could dive off and get into the water without being injured I could reach the shore by swimming.

As the sun dipped into the Pacific, I sneaked up behind the lookout man and gave him a tap over the right ear with the pin. He went to sleep instantly and I dived into the sea. After coming up and shaking the water from my head, I made for shore.

Fifteen or 20 minutes after I had crawled into the brush a commotion arose on board the freighter. Evidently I had been missed. I crawled further into the dense undergrowth and found a place that seemed to be rather safe. About an hour and a half later the boat weighed anchor and pulled out. I secured a couple of rocks and placed them against my body for use in case of emergency. Then exhaustion claimed me and I slept fitfully.

Day was breaking when I awoke and began to take stock of my surroundings. I was chilled to the bone. After about one hour of rubbing my limbs and rolling around, I decided to work my way inland. I had no idea of my whereabouts other than that I was sure I was yet on the Pacific coast of the United States. Being rather skilled in woodcraft, I started inland and eventually found a path leading into the interior. It did not matter to me where this path led; I meant to put as many miles between myself and that coastline as possible.

About 3:00 p. m. (as near as I could judge by observing the sun's position), I came upon a small cabin. Finding the cabin unlocked, I entered and from all appearances no one had been there for some time. However, I found pancake flour, lard, syrup, and coffee on a shelf and proceeded to prepare a much-needed meal. I ate and rested, then began to inspect the interior of the cabin. I found two pairs of socks under the crude bunk and were these welcome! My feet were blistered from walking. After a thorough examination of the quarters I decided to stay for the night.

At sun-up the next morning, I started trudging again. About noon the terrain began to slope inland and near 4:00 p. m. I came to a body of water. Offshore was a fishing launch and after considerable hallooing, I made the man understand what I wanted as he came in close enough to talk. After I convinced him that I was harmless, he took me aboard and landed me in Hoquiam, Washington. From there I made my way into Seattle. Securing some additional clothes, including an old rain coat, from a relief station of the Salvation Army, I rode freight trains until I reached Sandpoint, Idaho. Southeast of Sandpoint I found some wood cutting to do. After cutting and cording four cords of wood at $1.25 per cord, I wired my wife concerning my whereabouts. I had been skeptical about writing until then for I wanted to get inland as far as possible.

In Sandpoint, I met a couple of fellows who were going to the Boulder Dam job in Nevada. They insisted that work could be obtained there and I decided to accompany them. Returning to Spokane, we boarded a freight train on the Northern Pacific Railroad and made our way into Portland, Oregon. In Portland, we connected with the Southern Pacific Railroad and stayed with it until we reached Los Angeles. Living in Glendale, California, at that time was a man with whom I had been reared from boyhood. I stopped to see him. He was engaged in apian work. He tried to get me a job with his boss but failed. However, he staked me to $10.00.

Having lost the two men with whom I had made the trip south, I made the trip to Las Vegas and to Boulder City alone. No jobs were available so I waited until nightfall and boarded a Union Pacific freight train for Barstow, California. It was my intention to get back to the Southern Pacific road for it was rated among the "sun gazers" as the easiest one on the west coast to ride. I made a mistake and got out of Barstow on the Santa Fe Railroad. A brakeman chased me from the train in Bagdad, California, where I hunted for nearly an hour before I could find anyone who would even give me a drink of water.

That night I sneaked aboard another Santa Fe freight and arrived in Needles, California, the next morning. On the same train was a Mexican boy from Gallup, New Mexico. When he and I got off the train in Needles the sheriff took charge of us; putting us between the double tracks of the Santa Fe Railroad, he made us walk to Topock, Arizona. He would go toward Topock on one track and return on the other, changing his velocipede from one track to the other. The temperature was 116 degrees when we left Needles, and I think that was one of the hottest places into which I have ever gone, forcibly or of my own free will. When we arrived in Topock we were famished with hunger and almost dead of thirst. My feet were covered with blisters. Had it not been for the Mexican boy, I do not know what would have become of me. He left me lying under the west end of the Topock bridge with my shoes removed, bathing my feet to ease the pain.

About an hour later he returned with two big bowls of a Mexican dish and some bread. He had found friends among the section crew living there. He stayed with me until next day. I told him to go on, that I could make out some way. I punctured the blisters on my feet and continued the bathing treatment for two days and nights before I was able to wear my shoes. The third night I caught a freight train on the east end of the bridge and stayed on it until it reached Williams, Arizona. In Williams I picked up a truck ride to Phoenix, where I found friends and rested for a couple of days, giving my feet a chance to completely heal.

It was my intention upon leaving Phoenix to return home to El Paso, but when I reached Douglas I ran into "Arizona Jack" Harris. He and another friend were all packed up for a trip to Idaho for a try in the Forestry Service. The big fires were then raging in Idaho, northern California, and Washington. We left Douglas and went North in an old Jewett coach which belonged to Harris. It was equipped with hydraulic brakes, but they refused to work. Being somewhat of a mechanic I worked on the brakes several times on the route, but I could not get them to operate properly. We went from Gallup north through the Navajo Indian Reservation, and eventually to Weiser, Idaho.

In Weiser we tried to get into the Forestry Service, but they were not taking on men there. Near Weiser we passed through what was said to be the largest individually owned apple orchard in the world—10,000 acres of apple trees in one orchard. On our way we had crossed several mountain ranges and negotiated many difficult roads with no brakes on the car. We went through Spokane, then turned east to Sandpoint. We stopped in a tourist camp there for the night. The next morning when we got into the car to go into town to see about getting shipped out to the fires, the brakes on the Jewett started working perfectly.

About three days later we were signed into the Forestry Service and taken to Heron, Montana, via train. From Heron, we were taken to the base camp by truck. I had not passed through base camp until someone tacked onto me the nickname of "Texas Slim." This stuck during my association with the service until we separated on pay day at Heron.

From base camp we began the long hike to the forks of Ulm Creek in the Coeur d'Alene National Forest. On the hike we crossed a strip of burned forest. The ashes were shoe-mouth deep, and when we arrived to say we were dirty would be putting it mildly.

The second evening when we returned to camp there was quite a disturbance taking place at the entrance to the mess kitchen. The cook was referred to as the "grease ball" and one squint at him was convincing proof that the name had been properly applied. Red Ballard, a packtrain wrangler from Montana, had come in with his train. Red was about half starved as packers usually are at the end of the trail. Food in a Forestry Service camp is supposed to be ready and waiting for a packer when he has unloaded his string of pack animals and turned them to hay. No food was ready for Red and he was trying to get the "grease ball" to come outside and fight. The cook was backed up in the inclosure with a meat cleaver in his hand and telling Red, "You better stay out now," when several of us, including Mr. Jefferson, a chief ranger from Denver, walked up to the tables.

Mr. Jefferson took charge of the situation and turning to me said, "Slim, didn't you sign in as a cook?"

"Yes, sir," I replied, "and I can cook."

Turning to the "grease ball" he said, "Come out of that kitchen." "Get anyone you want to help you feed these packers," he said to me, "get them started away from here. Clean up this place and get some food prepared for these men."

Several flunkies were secured and put to work washing dishes, pots, pans, etc. We plunged into the work and in a few minutes sent Red and his partner away rejoicing. In about an hour and a half everything was spick and span and food was being prepared. This camp was prepared to feed and care for 225 men.

Breakfast and supper (or Dinnah, as it would be termed at tha Palais Royal or Delmonico's) were eaten in camp, but lunches had to be prepared for those who wended their way into the trenches each morning.

A Mr. Cherry from Missoula, Montana, had charge of the camp and commissary. I enjoyed many interesting conversations with him. Later, Chief Ranger Tom Crossley came in and assumed charge of the camp. He was still with us when the rains came and our camp was dismantled and supplies and equipment sent out.

When we broke camp and started out, the next camp up Ulm Creek was still in operation. The assistant cook and the "bull cook" wanted to go out so I took over the work of the "bull cook" until one of my partners came along on the way out. The camp was on the verge of breaking up anyway so I secured a release and came out. We were given our pay checks at Heron and furnished transportation to Sandpoint where we had signed up. After looking around for several days, we received the promise of a job cutting cord wood and I went up town to order a saw and some axes.

One of the old timers who had been out in the forest with us told me if I would go to Missoula with him we could get work cutting and burning dead and rotting timber from the national forests, which was done so that the bugs and borers would be destroyed before they could attack the live and growing specimens. The jobs did not mature, however, and in getting on a freight that night I got turned around and boarded one going east instead of west. I was homesick anyway, for when I came into Sandpoint from the fire a long letter from my wife was waiting for me and she was going to have a baby in October.

The next morning I was in Garrison, Montana. Between Missoula and Garrison several fellows who had come out of the forestry service had been hijacked and robbed on a train following the one on which I had ridden. From Garrison I rode the Chicago, Milwaukee, and St. Paul to Butte. In Butte, I found a friend who knew me by reputation. He sent me to a place where a meal could be obtained. I left out that night on the Oregon short line for Pocatello, Idaho, and Salt Lake City. In Salt Lake City I decided to return by way of San Francisco so I could again ride the Southern Pacific.

A group of hoboes, including myself, boarded a Union Pacific train which left Salt Lake just after dark. No one tried to stop us and we thought we were making progress when about five miles out the train came to a sudden stop and an automobile with glaring lights appeared on either side of the train. Five railroad bulls began to comb the train. After we were rounded up two of them drove the autos back to town while the other three walked the tracks and herded us the five miles back to the city. "Thought you were putting one over didn't you?" one of the bulls said. "Now listen," he warned, "if you fellows want to get out of Salt Lake City go down and ride the Denver & Rio Grande but don't get on the Union Pacific any more."

I do not know what the others did but I took him at his word. Catching a freight out over the Denver & Rio Grande Western, I came into Helper, Utah. I was hunting for a place to wash up and bathe my feet when I approached a ditch which ran along by a row of houses. I asked a man standing on a porch if the water in the ditch was all right to wash one's face and bathe one's feet in. "No," he answered, "that water contains alum. Come in and I'll give you some water in a tub." While I washed my face and hands and bathed my feet we became engaged in conversation which proved very interesting. He was a United Mine Worker and had been blacklisted for two years. He remembered seeing my name in the roster of the *United Mine*

*Workers' Journal* when I was secretary-treasurer of District No. 17. I spent a couple of hours with him, thanked and bade him good-bye.

I had sent most of the money drawn from the Forestry Service to my wife, keeping only a few dollars hidden away in my clothes for an emergency. I caught the next train out of Helper and hearing of John's Flophouse in Grand Junction, Colorado, where a cot and bath could be had for two bits, I decided to make that my next stop. Having secured a bath and a night's rest at John's, I went down to the yards next morning to catch a train. Before leaving the flophouse I had noticed two fellows eyeing me as though I was being appraised for some reason. While I was waiting for a train one of the same pair offered me an apple, which I declined.

Near the tracks was an old fellow whom I spotted right away as a professional hobo. He was cooking and making coffee.

"How goes it?" I greeted him.

"No good, any more," he growled, "what chance has a decent hobo got with all these punks riding the trains; none of them know from where they came or where they are going." Lowering his voice, he warned, "Look out for those two birds over there; they are no good." One of the two 'boes to which he referred was the one who had offered me the apple.

Soon a train came through and all boarded it. That night near Wolcott the pair who had been tagging me stuck up a group on the front end of the train and robbed them of sums ranging from 15 cents to $14.00. Kid Lewis, a wrestler from Lamar, Colorado, and his partner were in the group hijacked. After they had cleaned out the Kid, his partner, and about a dozen others, they came back over the train looking for me. I had gone toward the rear of the train, looking for an open reefer for the nights are cold crossing the Rockies. I had stopped within about 15 cars of the rear caboose when they came along.

They passed by me before I recognized who they were. Walking almost to the other end of the boxcar, they turned and came back toward me. The day before when I became suspicious of them I had secured a boulder about the size of a goose egg. I dropped it into the toe of a heavy cotton sock, tied a knot just above the boulder and another in the top end of the sock and had a pretty good improvised billy. When the pair came back to me one of them stayed on the walk board and the other stepped between me and the rear of the train.

"What do you want, fellows?" I inquired.

"We want your money and those shoes you have on," one of them growled.

I was standing sideways to him with a tight grip on the sock. I hit him along the side of the head, knocking him against his partner. While they were thus entangled I made a run for the rear end of the train. The moon was shining brightly and the train was barely crawling. The one whom I did not hit followed me to the rear end of the train and started to follow me off onto the ground. When I saw him start to climb down I began pelting him with rocks. He changed his mind and climbed back on top of the rear boxcar as the train crawled upgrade.

Finding no one at the depot in Wolcott, I crossed a small bridge to a little rooming house and applied for lodging. The people who ran the place looked at me inquiringly. "Yes, I got off that freight or was chased off by a couple of hijackers and if both of them aren't crazy or dope fiends I'm a Chinaman."

Several people exchanged glances and one of them remarked, "Blackie and his pard at it again, I'll bet." I described the pair as best I could and one of the lodgers spoke up, "That's them."

The next morning I caught a ride to Minturn on a truck. When I arrived there I ran into Kid Lewis, his pal, and the others. They had caught the pair of hijackers when they left the train at Minturn and had beaten them unmercifully. They were then arrested by a deputy sheriff who was stabbed through the hand with a knife while making the arrest. The sheriff arrived and took charge of the pair, thanking the Kid, his partner, and the others for assisting in their capture.

We were not disturbed again until we reached Colorado Springs. When we left the train there, we were rounded up and questioned concerning two convicts that had escaped from the Canon City penitentiary. They were described to us and we were asked if we had seen them. Giving satisfactory replies, we were permitted to continue our journey. In Pueblo, we were warned to be on the lookout for a bad railroad bull stationed at Trinidad. He met us when we got off the train, escorted us through the yards, and told us where to wait for the next train to pull out. I left the Kid and his partner and did not see them again.

I came through Dalhart, Texas, then turned west to El Paso and home. It was late at night when I arrived at the tents behind the Ascarate Drug Store into which we had moved before I was shanghaied. My wife told me she had searched the jails in El Paso and Juárez, Mexico, for me and that the police had radioed in all directions trying to locate me. She had also watched the papers for descriptions of the bodies that were occasionally fished from the Rio Grande River. During my absence neighbors and friends had aided and comforted her in her anxiety.

At that time El Paso County was doing some improvement work in McKelligan Canyon which was being converted into a public park. I secured work on this project which enabled me to rent a cabin in Camp Louise. We occupied it until our daughter was born.[1] When she was about three weeks old we moved into a cabin on the Bickley farm near Fabens, Texas. We started picking cotton, but a heavy snow fell, putting this business on the blink for almost two weeks. However, Mr. and Mrs. Bickley were nice people and helped us out in many ways. I secured another week's work on the Canyon project, riding 35 miles to and from work each night and morning in a truck. We stayed at the Bickley farm until about the middle of January, when the cotton crop was gathered and there was nothing more to do.

We then moved into a tourist camp facing the Ascarate Drug Store south of El Paso. I joined the El Paso County Political Alliance which was composed chiefly of the unemployed. We held some of our meetings in the coliseum and other times in lodge halls when they could be obtained. It was five miles from where I lived to the county courthouse, but I made the trip almost every day, continuously searching for employment. Sometimes I would return home for dinner, but more often a group of us would accept Sheriff Tom Armstrong's invitation to have lunch with the trusties in the El Paso County jail. Many times I saw him come down on the courthouse steps and tell the crowd that lunch was ready. Occasionally there would be 12 to 15 go to jail for a meal.

I was living in El Paso when the bonus marchers came through from California en route to Washington, D. C., and were joined by the El Paso contingent. The mayor of El Paso was a former railroad engineer. He used every influence at his command to relieve the suffering among the unemployed. Once it was reported that in reply to a communication sent to President Hoover by him asking help for the unemployed and starving Hoover offered the loan of machine guns, ammunition, and tear gas. It was reported that the mayor answered, "I do not believe the starving people would appreciate this kind of food." I participated in the nomination of Chris P. Fox for sheriff of El Paso County, but did not get to vote in the general election that fall.

Times became more difficult and work more scarce; we were compelled to move into the tents again. I leased a piece of ground from the owners of the Ascarate water works and set up the tents under the shade of some cottonwoods. My daughter was growing and for weeks at a time I did without smoking tobacco in order to be able to purchase graham crackers or some other tidbit for her. I wanted to keep her thriving. Up at 5:00 each morning, I watched every move from which a dime or a dollar could be made.

Many times about 5:00 a. m. I saw gunny sacks of Mexican liquor arrive on the American side. Wet labor was often used by some of the cotton planters. A group of wet Mexicans would come across, secure jobs, and work until the time arrived for them to be paid. When they demanded money the immigration authorities were tipped off. They were deported and the next night or two more wet labor would appear on the scene.

Early in September, a Mr. Hendricks came to me and asked me to go to Hill, New Mexico, and assist in building some houses for him. We left El Paso on September 12, 1932, and arrived on the Hendricks farm late at night. This farm, which was located three miles above Hill, was a part of the Dr. Hill ranch. Dr. Hill had migrated from Wheeling to this community years before. We stayed on the Hendricks farm until December 31, 1933. I assisted in building an adobe residence with 736 feet of floor space, two tenant houses (each with chicken houses and out buildings), a barn, a double garage, and a storage room.

Mr. Hendricks made me overseer of his farm shortly after I started working for him and from the farm in 1933 was marketed 128 bales of cotton. He and I disagreed soon after he moved in from El Paso in September, and on January 1, I moved to Radium Springs where I lived until we left for West Virginia on February 2, 1934.

Mr. David Williamson of the Maryland Coal Company of West Virginia sent for me to act as labor commissioner for him in conducting the receivership of that company. When I came to Wendel in Taylor County, the location of the Maryland Coal Company property, there were two unions functioning in the same mine. For about three months I dealt with a committee representing about 50 percent of the employees and another which spoke for the remainder of the miners. One half of the employees were members of an independent union chartered by the state, the others worked under a charter issued by the United Mine Workers of America. They were continuously quarreling and fighting among themselves.

One evening after having settled a grievance with both committees I said to Mr. Williamson, "Chief, how long are we going to keep this up?"

"Keep up what?" he inquired.

"This dealing with two separate committees."

"That's up to you," he said.

"Then I am going to put them all in one organization," I replied.

"Then inasmuch as we have a contract with the United Mine Workers, you had better have them all join that union," he said.

"Yes, that is what I had in mind," I agreed.

I held a meeting with the independents and told them what I would like for them to do. They were bitterly opposed to joining the United Mine Workers because of the officers then functioning in the district and local union. Some of the local officers at Wendel had been imported to break a strike several years before and one of them was an ex-convict. However, he had only broken a law for which few people had any respect anyway—the prohibition law. So, eventually, by much pleading and some coercion in a few individual instances, I had all the employees join the United Mine Workers of America. Thus were organized 173 men that the officials of the United Mine Workers had been trying to organize for years without success.

Numerous conferences were held among the miners after I came to Wendel as labor commissioner. Rumors of discussions about running me out on a certain date occasionally drifted through the mists of confusion. I only smiled and drove straight ahead to perfect my production organization. After the new agreement was signed in April, 1934, the man who had been functioning as assistant supervisor to Mr. Williamson was let go and I assumed full charge of operations. I adopted a policy of firmness yet did not permit anyone to act with more consideration toward me than I could show toward him. This policy paid dividends. Decent treatment will always bring out the best qualities in the average working man. He responds to fair treatment and if there is any good in him, proper consideration for his joys, his sorrows, his woes, his successes and failures, will develop that good.

One rule I established and rigidly enforced. No one in supervisory capacity was allowed to use profane language towards an employee. "If you do it," I warned, "head for the office when you finish for your money will be waiting for you." Relative to employees using profanity towards anyone in supervisory capacity the same rule was enforced. Sabotage was tried out on me several times when I first started, but I was able to recognize this practice immediately and they soon learned it was better not to try it.

When I returned to West Virginia in 1934, the United Mine Workers were in bad straits. A group had pulled away from the union and established the Progressive Miners Union. The internal wars, killings, and dynamiting that occurred are history. The hatreds and animosities, which are a direct result of the personal ambitions of a few men whose selfish regard for their own ego has worked innumerable hardships upon the membership of organized labor, have had to be dealt with.

It is a recognized fact that had it not been for the N.R.A., General Johnson, and his Blue Eagle, certain counties in southern West Virginia and other parts of the United States would yet be listed in the unorganized column, notwithstanding all statements of praise, whether they come from a labor leader, one of his "parrots," a United States Senator, or otherwise, to the contrary. In 1922, I made the remark to a prominent and influential man in Logan County, "Ten years from today a labor leader will be able to go into your county and hold a union meeting any time he chooses."

"You are crazy," he replied, laughing.

I just missed my guess by two years for by 1934 even the retail clerks and street sweepers were organized. The uprisings of 1919 and 1921 have often been referred to in calumny by those who wished to camouflage the issue or throw a smoke screen around their own activities. It was a known fact that for two years prior to 1921 the local unions of the Kanawha Valley, Paint Creek, Cabin Creek, and New River, were honeycombed with spies who advocated the purchase of guns, ammunition, and war supplies and constantly harangued the miners to march on Logan County. Many of these were weeded out and exposed but the "Miners' March" was a fruitful culmination of their activity. However, those who contributed vast sums of money to this undercover enterprise never dreamed that they were fomenting something that would destroy the very thing they were trying to protect, for that conflagration focussed the merciless eye of public opinion throughout the entire world on that intolerable situation.

When the miners surrendered their arms to Brigadier General H. H. Bandholtz in September, 1921, and said to the grizzled veteran of many battles, "General, we are not fighting our government," it was similar only to the signing of the Magna Carta by King John on the battlefield of Runnymede. Thus was established beyond question the fact that they were not in revolt against constituted authority but had taken to arms because they believed there was no other way to correct the wrongs perpetrated upon them by a conspiracy between the law and unlawful violence.

Today, we observe with misgivings the turmoil and strife which obtains in the ranks of labor; two great and powerful factions hurling epithets, howling charge and counter-charge at one another, while the sweeping arms of the Joy loader, the buckling head of the Morris Whaley, the revolving gatherers of the Sullivan and Jeffrey machines displace miners by the thousands. Every mechanical unit that is installed in the coal mines throws 40 to 50 men on the scrap heap and reclaims only eight to 12 of them for each seven-hour shift. Where do they and will they go? To Public Works Administration say some; on relief, say others. How long can this condition prevail, we ask?

# FOOTNOTES

### INTRODUCTION (Pages vii-xi)

1. Meticulous documentation of a narrative as condensed as this introduction would become absurd. Invaluable for the topic is Chapter XXXI of Charles Ambler and Festus P. Summers, *West Virginia, The Mountain State* (2d ed., Englewood Cliffs, New Jersey, 1958), 444-65. Other books cited below, the manuscript itself, and interviews with Mrs. Virginia R. Mooney provided essential information. On Mooney's death see Fairmont *Times*, February 25, 1952, and Fairmont *West Virginian*, February 25, 1952.

### CHAPTER I (Pages 1-9)

1. Also known as the Church of God in North America.
2. Chicago: International School for Social Economy, 1904. A tenth edition was published at Berkeley, California, by the same organization in 1914. Prominently identified with this school was George Gunton, editor of *Gunton's Magazine*.
3. Union organization advanced considerably in the Kanawha County mines during and after the coal strike of 1902, when President Theodore Roosevelt intervened in strikes centered in the anthracite fields of Pennsylvania. F. R. Dulles, *Labor in America, A History* (2d ed., New York, 1960), 188-193. The New River field in West Virginia was also partly organized at the same time, but only the Kanawha mines (except for Cabin Creek) remained organized after the crisis passed. W. P. Tams, Jr., *The Smokeless Coal Fields of West Virginia* (Morgantown, 1963), 42; Ambler and Summers, *West Virginia*, 444-47.
4. This camp was on Briar Creek, Washington District, Kanawha County, an unorganized area.
5. William S. Harris, *Hell Before Death* (Harrisburg, Pennsylvania, 1908). Published in 1907 under the title *Capital and Labor*.

### CHAPTER II (Pages 10-24)

1. A miner's "buddy" was the man who shared his "working place" in the mine. Tams, *Smokeless Coal Fields*, 35.
2. Cannelton.
3. The contract with the mines which were unionized expired on March 31, 1912, and the operators refused to renew it. Martial law was declared by Governor William E. Glasscock on September 2, 1912. J. R. Shanklin, plaintiff, *In the Circuit Court of Marshall County, West Virginia, In Re J. R. Shanklin. Habeas Corpus. Return of M. L. Brown, Warden of*

the *West Virginia Penitentiary, by Counsel* (Charleston, 1912), 4-5; subsequently cited as *In Re J. R. Shanklin*. Ambler and Summers, *West Virginia*, 448-49. For a summary of the situation and related documents see Elizabeth Cometti and Festus P. Summers, eds., *The Thirty-Fifth State; a Documentary History of West Virginia* (Morgantown, 1966), 522-33.

4. According to reports published in *In Re J. R. Shanklin*, 12, the strike was called on April 20, 1912.

5. See Chapter I, footnote 3.

6. This agency, to which there will be frequent reference in this narrative, also had offices in Thurmond and Bluefield, West Virginia, and in Denver, Colorado, and Richmond, Virginia. In 1912, their letterhead announced that they were special agents for the Norfolk & Western; Chesapeake & Ohio; Carolina, Clinchfield & Ohio; Virginian, and R. F. & P. railways. T. L. Felts to Justus Collins, Bluefield, November 7, 1912, Justus Collins Papers, West Virginia University Library; Edward Levinson, *I Break Strikes, the Technique of Pearl L. Bergoff* (New York, 1935), 26, 151-52, 209-10.

The operators claimed that there were few guards in the coal towns until the strike began. See Neil Robinson, *West Virginia on the Brink of a Labor Struggle* (Charleston, 1912), 6.

7. This incident occurred on August 30, 1912, and is elsewhere said to have taken place at Rhonda, a mining town near Dry Branch. See George S. Wallace, *In the Matter of the Hearing before a Sub-committee of the Committee on Education and Labor of the United States Senate* (Charleston, 1913), 3; *In Re J. R. Shanklin*, 46-49. The latter quotes one newspaper story describing Hodge as a mine check-weighman and another describing him as "a railroad man."

8. For a different version of this incident see *Ibid.*, 46-53.

9. September 2, 1912. The martial law zone included an area of Kanawha County roughly between Paint and Cabin creeks extending south to the Raleigh County line. *Ibid.*, 4-5.

10. Mary Jones, et al., appellants, *In the Supreme Court of Appeals of West Virginia, In Re Mary Jones, Charles H. Boswell, Charles Batley, Paul J. Paulson* (Charleston, 1913).

11. See Glasscock Papers, West Virginia University Library, for a copy of this report which reveals changes made after the first draft. Members of the Commission were Rt. Rev. P. J. Donahue, Wheeling; Captain S. L. Walker, Fayetteville; and Fred O. Blue, Charleston lawyer. The Commission mentioned as outstanding facts "the desperate efforts and often unwarranted and unlawful acts of the United miners to force the union into the disturbed districts, and the equally desperate, unwarranted and unlawful acts of the operators and their agents to keep the union out." They did not recognize a right to strike and were alarmed by the nature of union organizational efforts. *Report of West Virginia Mining Investigation Commission Appointed by Governor* [William E.] *Glasscock on the 28th Day of August, 1912* (Charleston, 1912), 10.

12. Mary "Mother" Jones, *Autobiography of Mother Jones,* ed. by Mary Field Parton (Chicago, 1925), 162, gives a slightly different version. For a sketch on Mother Jones and excerpts from her speeches see Cometti and Summers, *Thirty-Fifth State,* 529-33.

13. United States Congress, Senate, 63rd Cong., 1st Sess., Committee on Education and Labor, *Conditions in the Paint Creek District, West Virginia* (Washington, 1913), 3 vols. The committee concluded that actual peonage did not exist.

## CHAPTER III (Pages 25-38)

1. This chapter contains narratives of incidents between May, 1912, and April, 1913. Some which were in Mooney's manuscript have been omitted and the sequence sometimes has been established by internal evidence. Some duplication remains because most such incidents were similar in origin and detail. A summary of the events of this period will be found in Charles P. Anson, "A History of the Labor Movement in West Virginia," (Doctoral dissertation, University of North Carolina, 1940).

2. The first shooting at Mucklow occurred on May 28, 1912. *In Re J. R. Shanklin,* 9.

3. For a similar version see "Mother" Jones, *Autobiography,* 152-58.

4. Guards called miners Red Necks and strikers called strike breakers Red Necks.

5. For a different version of this incident see the Charleston *Gazette,* July 27, 1912. There was disagreement about the purpose of the guards and, as usual, about who started the shooting.

6. According to Kyle McCormick, *The New-Kanawha River and the Mine War of West Virginia* (Charleston, 1959), 139-40, this incident occurred in February, 1913. For a similar but much shorter version see "Mother" Jones, *Autobiography,* 161-62.

## CHAPTER IV (Pages 39-50)

1. William D. Haywood was leader of the Industrial Workers of the World, an industrial union which was organized in 1905. The I. W. W. spoke of class struggle and irrepressible conflict between the capitalist and the working class.

2. Evelyn L. K. Harris and Frank J. Krebs, *From Humble Beginnings, West Virginia State Federation of Labor, 1903-1957* (Charleston, 1960), 103-04. According to U. M. W. President John P. White's report when he retired in 1917, the International Executive Board suspended the autonomy of districts 17 and 29 in January, 1916. Autonomy was restored in 1917. *United Mine Workers' Journal,* 28, No. 29: 13, November 15, 1917.

CHAPTER V (Pages 51-62)
1. U. M. W. districts were reorganized following Keeney's revolt of 1915. District 17 at this time had headquarters in Charleston; it included the Kanawha area and all of West Virginia north of Charleston with the exception of eastern Preston County and the northern panhandle. Its jurisdiction also included unorganized southwestern portions of West Virginia and was extended in 1919 to include the Big Sandy and Elkhorn fields in eastern Kentucky. District 29, under Lawrence Dwyer, had headquarters at Beckley and included the New River and Winding Gulf fields of West Virginia. *United Mine Workers' Journal*, 30, No. 3: 11, February 1, 1919.
2. Red Warrior is actually on Tenmile Fork of Cabin Creek.
3. For a sketch of Clarence Watson see *Coal Age*, 45: 112, June, 1940; on Frank J. Hayes see *Coal Age*, 13: 876-77, May 11, 1918.
4. *United Mine Workers' Journal*, 29, No. 17: 4, September 15, 1918; *Ibid.*, 29, No. 18: 3, October 1, 1918.

CHAPTER VI (Pages 63-78)
1. Chapters VI and VIII through X deal with the United Mine Workers' effort to organize the southern West Virginia fields and related developments. There is little coverage of the overall situation, which Mooney apparently did not see. The southern fields were in competition with organized mines in Kanawha County, Pennsylvania, and the Middle West. Their organization was essential to a strong national union, but the operators in the area were openly resolved to resist union activity to the limit of their resources. Additional factors were the increased demand for coal preceding and during World War I, the insistence of the federal government on union recognition to prevent interruption of supply, and the postwar fluctuation in the demand and price for coal. Malcolm Ross, *Machine Age in the Hills* (New York, 1933), 142-43; James H. Thompson, *Significant Trends in the West Virginia Coal Industry, 1900-1957* (Morgantown, 1958), 6, 11, 16, 18; Ambler and Summers, *West Virginia*, 448-60.
2. Tams, *Smokeless Coal Fields*, 59-60. Chafin was elected sheriff in 1912 and served four years. In 1916, he was appointed to replace a man who had been elected county clerk but who resigned. His brother-in-law, Frank P. Hurst, was elected sheriff in 1916. Chafin was elected sheriff again in 1920 and served until convicted in 1924 of conspiring to violate U. S. liquor laws. George T. Swain, *The Incomparable Don Chafin* (Charleston, 1962), 4-7, 32.
3. *Report and Digest of Evidence Taken by Commission Appointed by Governor* [John J. Cornwell] *of West Virginia in Connection with the Logan County Situation* (Charleston, 1919), 24, 46-58; subsequently cited as *Report and Digest, Logan.*
4. For summaries of the following events from different points of view see McCormick, *The New-Kanawha River*, 144-46; George T. Swain, *Facts about the Two Armed Marches on Logan* (Charleston, 1962), 4ff; U. S.

Congress, Senate, 67th Cong., 1st Sess., Committee on Education and Labor, *West Virginia Coal Fields* (Washington, 1921), I, 51ff; subsequently cited as *West Virginia Coal Fields*.

5. The labor force of four leading companies in the area was reported to consist of 80 per cent "Americans" (50 per cent white, 30 per cent black) and 20 per cent "foreigners." *Report and Digest, Logan*, 39.

6. Some marchers returned home. An armed group of about 2,000 reportedly continued toward Logan (all figures on these events were given loosely and must be taken in the same manner). Governor Cornwell ultimately threatened to call for U. S. Army units, after which the marchers disbanded. Three special trains of the Chesapeake & Ohio Railway took them home on September 7, 1919. McCormick, *The New-Kanawha River*, 146; *United Mine Workers' Journal*, 30, No. 18: 22, September 15, 1919.

7. For a sketch on Tetlow see *Black Diamond*, 145: 18, December 3, 1960.

8. *West Virginia Coal Fields*, I, 103ff.

9. On Stanaford Mountain see above, Chapter II, 21. Ludlow, Colorado, was the site of a strike by the United Mine Workers against the Colorado Fuel and Iron Company in 1913. The strikers and their families were brutally suppressed by state militia. Dulles, *Labor in America*, 195-96; Samuel Yellen, *American Labor Struggles* (New York, 1936), 205-50.

10. See Winthrop D. Lane, "Labor Spy in West Virginia," *Survey*, 47: 10-12, October 22, 1921, which concerns Lively; *United Mine Workers' Journal*, 32, No. 15: 10, August 1, 1921; Lively's testimony before the Senate Committee on Education and Labor, *West Virginia Coal Fields*, I, 354-93.

11. This incident occurred on May 19, 1920. For other versions of the personalities and events see McCormick, *The New-Kanawha River*, 150-53; Ross, *Machine Age in the Hills*, 144; Winthrop D. Lane, *Civil War in West Virginia* (New York, 1921), 74-79. As Mooney relates later, he was not an eyewitness of these events but accepted versions reported by pro-Hatfield observers and probably Hatfield himself.

12. McCormick accepts the impression of Colonel Jackson Arnold, the first superintendent of West Virginia State Police, that Hatfield was joining the detectives to work in Beckley, and that the arrest was a cover which the townspeople did not understand, thus the shooting. McCormick, *The New-Kanawha River*, 152.

13. See Lane, *Civil War*, 74-79. According to other accounts, there would have been a longer interval between these reconstructed exchanges between Hatfield, Testerman, and the detectives than this narrative allows. Hatfield told the U. S. Senate committee which investigated in 1921 that he had not asked anyone for help. McCormick, *The New-Kanawha River*, 151-53; *United Mine Workers' Journal*, 31, No. 13: 8, July 1, 1920; *West Virginia Coal Fields*, I, 206.

14. According to his testimony before the U. S. Senate committee, Mooney was at his residence when informed by telephone of the Matewan

massacre. But Lively, in his testimony, told the committee that Mooney was informed of the shooting at union headquarters. *Ibid.*, I, 16, 363-64.
15. *United Mine Workers' Journal*, 32, No. 7: 13, April 1, 1921; Virgil Carrington Jones, *The Hatfields and the McCoys* (Chapel Hill, 1948), 238.
16. *United Mine Workers' Journal*, 32, No. 14: 3, July 15, 1921; *West Virginia Coal Fields*, I, 166-67.

CHAPTER VII (Pages 79-85)

1. "Mother" Jones, *Autobiography*, 238.
2. General Alvaro Obregón had been elected president on September 5, 1920, following the forcible overthrow of President Venustiano Carranza, who was killed in renewed military conflict. Obregón was not recognized by the United States until August, 1923, after agreeing to respect titles to land acquired by citizens of the United States before 1917 and to adjust United States claims.
3. Thomas J. Mooney was a member of the moulders' union who in 1916 was attempting to organize street car workers. Tom Mooney and Warren Billings were convicted in connection with an explosion during a "Preparedness Parade" in San Francisco in July, 1916. The "Mooney-Billings case" became a cause célèbre of the labor and protest movements of the day. Curt Gentry, *Frame-Up; the Incredible Case of Tom Mooney and Warren Billings* (New York, 1967); Elizabeth Gurley Flynn, *I Speak My Own Piece; Autobiography of "The Rebel Girl"* (New York, 1955), 204-05.
4. *Confederatión Regional Obrera Mexicana* (Confederation of Mexican Labor), formed in 1918, comparable to the American Federation of Labor. It played a significant role in the Mexican reform movement, although it was soon challenged by a more radical organization.
5. Eugène Sue was the pen name of Marie Joseph Sue (1804-57), French novelist who was strongly affected by the socialist ideas of his day. After the revolution of 1848 he sat in the national assembly, but he was exiled in 1851 for political reasons. Mooney occasionally paraphrased from English translations of Sue's works.
6. Pseudonym of Arthur Desmond, novelist at the turn of the century.
7. Mooney reported on his Mexican trip in the *United Mine Workers' Journal*, 32, No. 7: 12, April 1, 1921. He described the revolutionary government as "one of the most liberal and friendly . . . towards the workers that is in existence. . . ." and called for its recognition by the United States.

CHAPTER VIII (Pages 86-100)

1. Governor of West Virginia, 1921-25, was E. F. Morgan, Republican from Fairmont. The strike-lockout in Mingo County continued through the winter of 1920-21 and into the summer. In December, 1920, about 400 families, or over 10,000 persons, were supported by the U. M. W. in tent colonies in Lick Creek, Nolan, Thacker, Lind, Sprigg, Naugatuck, and

Blocten. *United Mine Workers' Journal,* 32, No. 2: 14, January 15, 1921.
2. Hatfield was charged with inciting a shooting in Mohawk, a town in McDowell County. Ambler and Summers, *West Virginia,* 456.
3. Kirkpatrick was a deputy sheriff. McCormick, *The New-Kanawha River,* 154. In the following paragraphs, Mooney reconstructs events from the reports of others; as he mentions earlier, he was in Washington at this time. Mrs. Chambers testified that she and her husband were in front of the Hatfields. *West Virginia Coal Fields,* II, 741.
4. *United Mine Workers' Journal,* 32, No. 16: 3, August 15, 1921; *Ibid.,* 32, No. 17: 10, September 1, 1921. Lively was indicted and tried. He admitted killing Hatfield but claimed self-defense because Hatfield had shot first; he was acquitted.
5. *Ibid.;* Swain, *Facts about the Two Armed Marches,* 12-13.
6. Drawdy Creek.
7. Probably a very exaggerated figure. See Ambler and Summers, *West Virginia,* 457; McCormick, *The New-Kanawha River,* 158.
8. The indictments against Keeney, Mooney, Blizzard and others were returned by grand juries which sat in Logan County in September and October, 1921, and January, 1922; and in Kanawha County in October, 1921. Most of the later proceedings were a result of changes of venue requested by the defense. See Miners' Treason Trial Papers, A & M 979, Reel 1, West Virginia University Library.

## CHAPTER IX (Pages 101-18)

1. Keeney told a similar story. Swain, *Facts about the Two Armed Marches,* 36-37.

## CHAPTER X (Pages 119-30)

1. John L. Lewis, "A Union's Non-Union Mines," *Nation,* 120: 287-88, March 18, 1925.
2. Lewisburg was seat of Greenbrier County, to which changes of venue had been granted.
3. Tetlow and Van A. Bittner later testified before the U. S. Senate Committee on Interstate Commerce that the international union took over the West Virginia organization at the request of the Northern West Virginia Coal Operators Association. Under the existing contract the international organization was responsible for enforcement of the agreement. Tetlow had been in West Virginia since 1923 at the request of the district organization. United States Congress, Senate, 70th Cong., 1st Sess., Committee on Interstate Commerce, *Conditions in the Coal Fields of Pennsylvania, West Virginia, and Ohio, Hearings* . . . (Washington, 1928), I, 1128-29, 1420-24; Anson, "History of the Labor Movement in West Virginia," 127.
4. Bittner and other union leaders believed that the breaking of union agreements by the New River operators in 1922 was the beginning of an

anti-union movement that spread throughout West Virginia and most of the nation. West Virginia was almost completely de-unionized by 1928. *Ibid.*, 237; *Conditions in the Coal Fields*, II, 1445.

### CHAPTER XI (Pages 131-41)

1. See "Mother" Jones, *Autobiography*, 137-39, 172-77.

### CHAPTER XII (Pages 142-49)

1. Keeney had been a delegate to the March 10, 1930, convention in Springfield, Illinois, at which the anti-Lewis organization claimed to have become the legally constituted United Mine Workers of America. They were commonly referred to as the Reorganized. Irving Bernstein, *A History of the American Worker, 1920-1933; the Lean Years* (Boston, 1960), 366-77. At this time there were less than 600 U. M. W. members in West Virginia. H. L. Morris, *The Plight of the Coal Miner* (Philadelphia, 1934), 126.

2. Howatt was a leader of Kansas District 14, U. M. W.; Walker was a former miner who was at this time president of the Illinois Federation of Labor; Germer had been secretary of the Socialist Party but was at this time a real estate agent. Also associated with the anti-Lewis movement was Oscar Ameringer, Socialist publisher of the *Illinois Miner*. Bernstein, *History of the American Worker*, 368.

3. Bittner was president of U. M. W. District 5, 1911-16, and an organizer, international representative, and committeeman for the U. M. W. and C. I. O.

4. Lewis had effectively crushed this revolt by the summer of 1930, gaining control in Illinois, the center of the movement. The decision referred to was from the Lee County Circuit Court of Dixon, Illinois, which restored the Lewis organization to the leadership of District 12 in Illinois. The Springfield convention was declared to have had no authority to reorganize the U. M. W. Actually, Lewis did not gain complete control of District 12 until February, 1933, and even then a rival organization, the Progressive Miners, held out with headquarters at Gillespie, Illinois. Bernstein, *A History of the American Worker*, 372-77.

### CHAPTER XIII (Pages 150-65)

1. According to family records, a daughter was born on October 25, 1931, at El Paso.

## BIBLIOGRAPHY

This bibliography is an effort to list some of the printed materials most useful for the period covered by Mooney's narrative. No attempt is made to present an exhaustive listing on the history of coal mining and labor in West Virginia. For comprehensive coverage of all aspects of the industry the scholar should consult Robert F. Munn, *The Coal Industry in America, a Bibliography and Guide to Studies* (Morgantown, 1965), which includes West Virginia references not listed here.

References to West Virginia events in the *United Mine Workers' Journal* are so frequent during these years that they are not listed as separate articles. There are footnote references to the *Journal* where confirmation or additional information was regarded as useful.

Relevant manuscript holdings are listed and indexed in *Guide to Manuscripts and Archives in the West Virginia Collection* (No. I, Morgantown, 1958, by Charles Shetler, and No. II, Morgantown, 1965, by F. Gerald Ham). Particularly relevant are the papers of governors William E. Glasscock, Henry D. Hatfield, and John J. Cornwell, and the papers of Justus Collins, a southern West Virginia coal operator.

### PRINTED MATERIALS

Ambler, Charles, and Festus P. Summers. *West Virginia, the Mountain State.* 2d ed. Englewood Cliffs, New Jersey, 1958.
    Chapter XXXI, "Labor Moves Forward," includes an excellent summary of the period covered by Mooney.
American Constitutional Association. *Life in a West Virginia Coal Field.* Charleston, 1923.
    A rosy description from the viewpoint of the operators.
"Anthracite Coal Crisis and Conditions in West Virginia." *Outlook*, 82: 575-78, March 17, 1906.
Baltimore *Sun*. *Mingo County West Virginia Coal Strike.* Baltimore, 1921.
    Reprint of a series of articles, January 23-25, 1921.
Bernstein, Irving. *A History of the American Worker, 1920-33; the Lean Years.* Boston, 1960.
    Particularly worthwhile on the conflicts within the U. M. W. on the national level.
Carter, Charles F. "Murder to Maintain Coal Monopoly." *Current History*, 15: 597-603, January, 1922.
—————. "The West Virginia Coal Insurrection." *North American Review*, 198: 457-69, October, 1913.
Cartlidge, Oscar. *Fifty Years of Coal Mining.* Charleston, 1936.
"Clarence Watson." *Coal Age*, 45: 112, June, 1940.
Coleman, McAllister. *Men and Coal.* New York, 1943.

———————. "A Week in West Virginia." *Survey*, 53: 532-34, February 1, 1925.

Cometti, Elizabeth, and Festus P. Summers, eds. *The Thirty-Fifth State; a Documentary History of West Virginia*. Morgantown, 1966. Contains important documents relating to labor problems and excellent summaries of the history of the period.

Conley, Phil. *History of the West Virginia Coal Industry*. Charleston, 1960.

Davis, Jerome. "Human Rights and Coal." *Journal of Social Forces*, 3: 102-6, November, 1924.

Dulles, Foster R. *Labor in America*. 2d ed., New York, 1960.

Eavenson, Howard N. *The First Century and a Quarter of American Coal Industry*. Pittsburgh, 1942.

Emmet, Boris. *Labor Relations in the Fairmont, West Virginia, Bituminous Coal Field*. Washington, 1924. (United States Department of Labor, Bureau of Labor Statistics, Bulletin No. 361) A worthwhile report on relationships between the union and the operators at a time when the relationship was changing rapidly.

Evans, Chris. *History of the United Mine Workers, 1860-1900*. Indianapolis, 1918. 2 vols.

Fisher, W. E. and Anne Bezanson. *Wage Rates in the Bituminous Coal Industry*. Philadelphia, 1932.

Flynn, Elizabeth Gurley. *I Speak My Own Piece; Autobiography of "The Rebel Girl"*. New York, 1955.

"Frank J. Hayes." *Coal Age*, 13: 876-77, May 11, 1918.

Frankfurter, Felix, and Nathan Greene. *The Labor Injunction*. New York, 1930 (Reprinted, Gloucester, Massachusetts, 1963). Covers legal and social history of the Hitchman case and similar judicial decisions which were important for the West Virginia situation.

Gentry, Curt. *Frame-Up. The Incredible Case of Tom Mooney and Warren Billings*. New York, 1967. Careful study of a case which inspired much emotion among the mine workers.

Gleason, Arthur. "Company-owned Americans." *Nation*, 110: 794-95, June 12, 1920.

———————. "Private Ownership of Public Officials." *Nation*, 110: 724-25, May 29, 1920.

Glueck, Elsie. *John Mitchell, Miner*. New York, 1929.

Hall, R. Dawson. "The Fairmont, West Virginia, Coal Region." *Coal Age*, 1: 138-43, November 11, 1911.

Harris, Evelyn L. K., and Frank J. Krebs. *One Hundred Years with West Virginia Labor*. Charleston, 1963.

———————. *From Humble Beginnings, West Virginia State Federation of Labor, 1903-1957*. Charleston, 1960.

Hinds, Roy W. "The Last Stand of the Open Shop." *Coal Age*, 18: 1037-40, November 18, 1920.

Hinrichs, A. F. *The United Mine Workers of America and the Non-Union Coal Fields*. New York, 1923. (Columbia University Studies, CX, 1923)

Huberman, Leo. *The Labor Spy Racket.* New York, 1937.

Hudson, Harriet D. *The Progressive Mine Workers of America; a Study in Rival Unionism.* Urbana, Illinois, 1952. (Bureau of Economic and Business Research. Bulletin 73)

Hunt, Edward E., et al. *What the Coal Commission Found; an Authoritative Summary by the Staff.* Baltimore, 1925.

Jones, Mary, et al. *In the Supreme Court of Appeals of West Virginia, In Re Mary Jones, Charles H. Boswell, Charles Batley, Paul J. Paulson.* Charleston, 1913.

Jones, Mary. *Autobiography of Mother Jones.* ed. by Mary Field Parton. Chicago, 1925.

Jones, Virgil Carrington. *The Hatfields and the McCoys.* Chapel Hill, 1948.
A colorful account of the famous feud which describes the transformations wrought by the coal economy but which overemphasizes the relation of the feud to the union conflicts.

Jordan, Margaret W. "A Plea for the West Virginia Miner." *Coal Age,* 6: 914-16, December 5, 1914.

Kirchwey, Freda. "Miners' Wives in the Coal Strike." *Century,* 105: 83-90, November, 1922.

Lambie, Joseph T. *From Mine to Market; the History of Coal Transportation on the Norfolk and Western Railroad.* New York, 1954.

Lane, Winthrop D. *Civil War in West Virginia; a Story of the Industrial Conflict in the Coal Mines.* New York, 1921.
A colorful account from the miner's point of view, 1919-1921, written first as a series of newspaper articles for a New York paper.

——————. *The Denial of Civil Liberties in the Coal Fields.* New York, 1924.

——————. "Labor Spy in West Virginia." *Survey,* 47: 110-12, October 22, 1921.

Levinson, Edward L. *"I Break Strikes!" The Technique of Pearl L. Bergoff.* New York, 1935.
Contains harsh references to the Baldwin-Felts detectives.

Lewis, John L. *The Miners' Fight for American Standards.* Indianapolis, 1925.

——————. "A Union's Non-union Mines." *Nation,* 120: 287-88, March 18, 1925.

Lynch, Lawrence R. "The West Virginia Coal Strike." *Political Science Quarterly,* 29: 626-63, December, 1914.

McCormick, Kyle. *The New-Kanawha River and the Mine War of West Virginia.* Charleston, 1959.
A careful and restrained coverage of the leading events.

——————. "The National Guard of West Virginia during the Strike Period of 1912-1913." *West Virginia History,* 22: 34-35, October, 1960.

McGill, Nettie P. *Welfare of Children in the Bituminous Coal Mining Communities in West Virginia.* Washington, 1923. (Children's Bureau Publication 117)

Mathews, William G. *Martial Law in West Virginia, an Address.* Washington, 1913. (United States Congress. Senate. 63d Cong. 1st Sess. Senate Document 230)

Michelson, M. "Feudalism and Civil War in the United States of America." *Everybody's Magazine,* 28: 615-28, May, 1913.

Morris, Homer L. *The Plight of the Coal Miner.* Philadelphia, 1934.

Northern West Virginia Coal Operators Association. *The Coal Industry of the State of West Virginia.* Fairmont, 1923.

Older, Cora (Mrs. Fremont). "Answering a Question: Martial Law in West Virginia." *Colliers,* 51: 26, 28, April 19, 1913.

—————. "Last Day of the Paint Creek Court Martial." *Independent,* 74: 1085-88, May 8, 1913.

Owens, John W. "Gunmen in West Virginia." *New Republic,* 28: 90-92, September 21, 1921.

Parsons, Floyd W. "Mining Coal on the Virginia Railroad." *Coal Age,* 1: 1039-43, May 18, 1912.

"Percy Tetlow." *Black Diamond,* 145: 18, December 3, 1960.

Pocahontas Operators Association. *The Bill, the Answer, the Injunction in the Case of the Pocahontas Operators Association against Organizers, Agitators and Others Having Designs on Interfering with a Non-Union Labor Situation.* Bluefield, 1921.

A clear expression of the non-union operator's intent and attitude.

Randall, James G. "Miners and the Law of Treason." *North American Review,* 216: 312-22, September, 1922.

Red Jacket Consolidated Coal and Coke Company, et al. *Transcript of the Record. District Court of the United States for the Southern District of West Virginia . . . Red Jacket, et al., plaintiffs, versus John L. Lewis, et al.* Charleston [1923]. 3 vols.

*Report and Digest of Evidence Taken by Commission Appointed by the Governor of West Virginia* [John J. Cornwell] *in Connection with the Logan County Situation.* Charleston, 1919.

*Report of West Virginia Mining Investigation Commission Appointed by Governor* [William E.] *Glasscock on the 28th Day of August, 1912.* Charleston, 1912.

Robinson, Neil. *West Virginia on the Brink of a Labor Struggle.* Charleston, 1912.

Rochester, Anna. *Labor and Coal.* New York, 1931.

Contains a brief but vigorous summary of the conflict from 1912 to 1921.

Ross, Malcolm. *Machine Age in the Hills.* New York, 1933.

Roy, Andrew. *History of the Coal Miners of the United States, from the Development of the Mines to the Close of the Anthracite Strike of 1902.* 3d ed. Columbus, Ohio, 1907.

Shanklin, J. R., plaintiff. *In the Circuit Court of Marshall County, West Virginia. In Re J. R. Shanklin. Habeas Corpus. Return of M. L. Brown,*

*Warden of the West Virginia Penitentiary.* Charleston, 1912.

Includes a series of reports and official declarations relating to the incidents between April 1 and October, 1912.

Shepherd, William G. "Big Black Spot." *Colliers,* 88: 12-13, September 19, 1931.

"Sociological Conditions in West Virginia." *Coal Age,* 2: 733-34, November 23, 1912.

Spivak, John L. *A Man in His Time.* New York, 1967.

The author was a reporter who covered the conflict in West Virginia.

*Statement to the United States Coal Commission by Non-Union Operators of Southern West Virginia.* Bluefield, 1923.

Suffern, Arthur E. *The Coal Miners' Struggle for Industrial Status.* New York, 1926.

No detail on West Virginia but good on the broader problem.

—————. *Conciliation and Arbitration in the Coal Industry of America.* Boston and New York, 1915.

Includes a special consideration of the West Virginia situation as a problem for the industry.

Swain, George T. *Facts about the Two Armed Marches on Logan.* Charleston, 1962.

—————. *The Incomparable Don Chafin: Review of the Life of Logan's Dauntless and Indomitable Sheriff, Who Prevented the Invasion of Logan County on Two Occasions by Armed Miners from the Kanawha Valley Coal Fields.* Charleston, 1962.

Tams, W. P., Jr. *The Smokeless Coal Fields of West Virginia.* Morgantown, 1963.

The best single source on the area by one of the leading operators.

Thompson, James H. *Significant Trends in the West Virginia Coal Industry, 1900-1957.* Morgantown, 1958. (West Virginia University, Business and Economic Studies, Vol. 6, No. 1)

Thurmond, Walter. *The Logan Coal Field of West Virginia; a Brief History.* Morgantown, 1964.

The author was one of the area's leading operators and for many years secretary of the Southern Coal Producers Association.

United Mine Workers of America. *Memorandum of Agreement Made and Entered into this 28th Day of March, 1924, by the Membership of the United Mine Workers of America and the Membership of the Northern West Virginia Coal Operators Association.* Fairmont, 1924.

—————. *Proceedings of the Convention of the United Mine Workers of America.*

—————. *John L. Lewis and the International Union; United Mine Workers of America, the Story from 1917 to 1952.* Washington, 1952.

United States Congress. Senate. 63rd Cong. 1st Sess. Committee on Education and Labor. *Conditions in the Paint Creek District, West Virginia.* Washington, 1913. 3 vols.

—————. —————. 63rd Cong. 2d Sess. Committee on Education and Labor. *Investigation of Paint Creek Coal Fields of West Virginia . . . Report.* Washington, 1914.

—————. —————. 67th Cong. 1st Sess. Committee on Education and Labor. *West Virginia Coal Fields; Hearings . . . to investigate the recent acts of violence in the coal fields of West Virginia and adjacent territory and the causes which led to the conditions which now exist in said territory.* Washington, 1921-22. 2 vols.

—————. —————. 70th Cong. 1st Sess. Committee on Interstate Commerce. *Conditions in the Coal Fields of Pennsylvania, West Virginia, and Ohio. Hearings . . .* Washington, 1928. 2 vols.

Wallace, George S. *In the Matter of the Hearing before a Subcommittee of the Committee on Education and Labor of the U. S. Senate.* Charleston, 1913.

Warner, Arthur. "West Virginia—Industrialism Gone Mad." *Nation*, 113: 372-73, October 5, 1921.

West, Harold E. "Civil War in the West Virginia Coal Mines." *Survey*, 30: 37-50, April, 1913.

West Virginia. Department of Labor. *Biennial Reports.*

—————. Department of Mines. *Annual Report.*

—————. State Federation of Labor. *Proceedings.*

Wilson, Edmund. "Frank Keeney's Coal Diggers." *New Republic*, 67: 195-99, 229-31, July 8, 15, 1931.

Yellen, Samuel. *American Labor Struggles.* New York, 1936.

Contains a detailed account of the "Ludlow massacre."

## THESES AND DISSERTATIONS

Anson, Charles P. "A History of the Labor Movement in West Virginia." Doctoral dissertation, University of North Carolina, Chapel Hill, 1940.

Barb, John M. "Strikes in the Southern West Virginia Coal Fields, 1912-1922." Thesis, West Virginia University, Morgantown, 1949.

Campbell, Roy E. "History of the Development of the Coal Industry in Kanawha District, West Virginia." Thesis, West Virginia University, 1930.

Crawford, Charles B. "The Mine War on Cabin Creek and Paint Creek, West Virginia, in 1912-13." Thesis, University of Kentucky, 1939.

Hall, Betty Snyder. "The Role of Rhetoric in the Northern West Virginia Activities of the United Mine Workers, 1897-1927." Thesis, West Virginia University, 1955.

Harvey, Helen B. "From Frontier to Mining Town in Logan County, West Virginia." Thesis, University of Kentucky, 1942.

Hurst, Mary B. "Social History of Logan County, West Virginia, 1765-1923." Thesis, Columbia University, 1924.

Merrill, William M. "Economics of the Southern Smokeless Coals." Doctoral dissertation, University of Illinois, 1953.

Posey, Thomas E. "The Labor Movement in West Virginia." Doctoral dissertation, University of Wisconsin, 1948.
Trail, William R. "History of the United Mine Workers in West Virginia, 1920-1945." Thesis, New York University, 1950.
White, Elizabeth. "Development of the Bituminous Coal Mining Industry in Logan County, West Virginia." Thesis, Marshall College, 1956.

## INDEX

Abilene, Texas, 131
Agua Caliente, Arizona, 133
Aiello, Nick, 53, 55
Ajo, Arizona, 138, 139
Akers, Tom, 68, 69
Alabama, 30
Alaska, 151
Albuquerque, New Mexico, 131
Allen, Walter, 123
American Correspondence School of Law, 47
American Federation of Labor, 80
American Legion, 123
American Plan, ix
American Rifle Association, 29
Ameringer, Oscar, 173, Ch. XII, note 2
Anti-Catholicism, 48-50
Appalachia, viii
Arizona, x, 131-35, 137-41, 154
Arizona Biltmore, 137-38
Armenia, 60
Armistice Day, 77-78
Armstrong, Tom, sheriff of El Paso County, Texas, 160-61
Arnold, Col. Jackson, 170, Ch. VI, note 12
*Art of Lecturing*, 8
Ascarate Drug Store, 160
Ash, Nathan, 57, 88
Associated Press, 76, 86, 94
Athens, Ohio, 123
Aurora, Missouri, 48
Avis, S. B., lawyer, 115

Bad Eye, 29
Bagdad, California, 153
Bailey, Judge R. D., Mingo County, 101
Baldwin-Felts Detectives, 15, 72-76, 103, 115, 167, Ch. II, note 6
Ballard, Red, 155
Baltimore, Maryland, 122, 124-25
Balzac, 48
Bandholtz, Brig. Gen. H. H., 91-93, 95, 98, 164
Baptists and baptism, 1-3
Barboursville, West Virginia, 64
Barstow, California, 153
Batley, C. H., 46
Battle of Blair Mountain, 95-98
Beatty, Nelson, U. M. W. leader, 143
Beckley, West Virginia, 169, Ch. V, note 1; 170, Ch. VI, note 12
Belcher, A. M., Charleston lawyer, 120, 122
Berkeley Springs, West Virginia, 123
Bernarding's saloon, 43
Bible, 7

*Bible Scenes and Studies*, 7
Big Sandy coal field, Kentucky, 169, Ch. V, note 1
Billings, Warren, 171, Ch. VII, note 3
Bittner, Van A., U. M. W. leader, 143, 146-48; 172, Ch. X, notes 3, 4; 173, Ch. XII, note 3
Bituminous Coal Commission, 1919, 67
Black Betsy, West Virginia, 44-45, 56
Blacklist, 12, 46-47, 157
Blair, West Virginia, 93-98
Blair Mountain, 95-98
Bland, Judge Robert, Logan County, 97, 115, 118
Blizzard, William, U. M. W. leader, Illustrations, 67, 69-72, 90, 98, 107-08, 112, 116-18, 121-23, 126, 130
Blocten, West Virginia, 172, Ch. VIII, note 1
Blue, Fred O., Charleston lawyer, 167, Ch. II, note 11
Blue Eagle, N. R. A., 128, 164
Bluefield, West Virginia, 167, Ch. II, note 6
Boggs, A. H., miner, 101
Bolsheviks or Bolshevist, 40, 117
Bombs, 97-98, 144
Bonaparte, Napoleon, 60
Bonus marchers, 161
Booher, detective, 73
Boomer, West Virginia, 53
Boone County, West Virginia, 67, 90, 119
Borah, Senator William E., Idaho, 23
Boston, Massachusetts, 22, 88
Boulder, Colorado, 153
Boulder Dam, 153
Bower, West Virginia, 56
Braxton County, West Virginia, 56
Breedlove, Alex, miner, 78
Briar Creek, Kanawha County, West Virginia, 166, Ch. I, note 4
Brockus, Capt. J. R., West Virginia State Police, 95
Brown, John, 121
Brown, Mr., 147
Brownsville, Texas, 29
"Buddy", 12, 166, Ch. II, note 1
Bullethead, 29
Bull Moose Special, 35-38
Burns, W. J., detectives, 120
Butte, Montana, 157

Cabin Creek, West Virginia, viii, 14-39, 45, 55, 112, 164, 166, Ch. I, note 3; 169, Ch. V, note 2
Cabin Creek Junction, West Virginia, 30, 95, 142
Cairns, Thomas, U. M. W. leader, 41-45
California, 153-54, 157
Calumet Arizona Copper Company, 138, 141
Camp Dix, New Jersey, 98
Camp Montezuma, Arizona, 132
Camp Taylor, 98
Canada, 126
Cannelton, West Virginia, 15, 39, 44, 47-48, 166, Ch. II, note 2
Canon City, Colorado, 159
Carlsbad, New Mexico, 137
Carlyle, Thomas, 48

Carmichael, Mr., 147
Carmichael Auditorium, 147-48
Carolina, Clinchfield, and Ohio Railroad, 167, Ch. II, note 6
Carr, George, miner, 142
Carranza, President Venustiano, 171, Ch. VII, note 2
Catholics, 49
Caverna, Hotel, Carlsbad, New Mexico, 137
Central Competitive Field, vii-viii
Chaffin, Pleas, 101-02
Chafin, "Con", prosecuting attorney, Logan County, West Virginia, 115
Chafin, Don, sheriff of Logan County, 64, 94, 97, 99, 102, 104, 106, 110, 115-16, 118, 120, 169, Ch. VI, note 2
Chambers, "Daddy" Reese, 75
Chambers, Ed, 74-76, 87-89, 99, 103, 108, 172, Ch. VIII, note 3
Chambers, Mrs. Ed, 88, 172, Ch. VIII, note 3
Change of venue, 104, 116, 119, 122-23
Chaplin, Ralph, miner, 38
Chappel, Jim, 133
Charleston, West Virginia, vii, ix-x, 39, 52, 64, 66-67, 69, 71, 76, 79, 91, 95-96, 99-100, 106, 118-19, 122-24, 128, 130, 135-36, 142, 169, Ch. V, note 1
Charleston (West Virginia) *Daily Gazette*, 116
Charles Town, West Virginia, 120, 122-23, 126
Cherry, Mr., 156
Chesapeake and Ohio Railroad, 30, 35, 96-97, 110, 135, 167, Ch. II, note 6; 170, Ch. VI, note 6
Chicago, Illinois, 88
Chicago, Milwaukee, and St. Paul Railroad, 157
Christian, George B., secretary to President Warren G. Harding, 91
Church of God in North America, 166, Ch. I, note 1
Cincinnati, Ohio, 22, 88, 122
Clarksburg, West Virginia, 58-59, 142, 145, 147
Clearwater, Florida, 131
Clearwater (Florida) Building Trades Council, x
Clendenin, Obe, 51
Cleveland, Ohio, 119
Cline, Lawyer, 101
Coal companies, SEE Coal operators
Coal operators, vii-viii, 15-16, 18, 22-23, 27, 34-35, 39, 41, 43-44, 46, 58-60, 63-64, 69-71, 86, 88-90, 124-27, 147, 162
Coal River, West Virginia, 67, 95, 119
Coates, Robert, detective, 120
Coeur d'Alene National Forest, Idaho, 155
Collins, John, 87-88
Colorado, 43, 158-59
Colorado Fuel and Iron Company, 170, Ch. VI, note 9
Colorado Springs, Colorado, 159
Columbia, Missouri, 136
Columbus, Ohio, 99
Communists, 122
Company guards, 12-18, 21, 25-29, 57, 59-60, 64-66, 72-76, 86-88, 95, 97, 101, 103, 120-21, 129; SEE ALSO deputy sheriffs and gunmen
Company store, 11, 47
Confederatión Regional Obrera Mexicana (C. R. O. M.), 82, 171, Ch. 7, note 4
Consolidated Coal Company, 147
Consolidation Coal Company, 60

Coolidge, President Calvin, 129
Copen, West Virginia, 56
Copenhaver, John, deputy sheriff, Kanawha County, 100
Copper Queen, 137, 140
Cornwell, Governor John J., West Virginia (1917-21), 59-60, 63-66, 129, 170, Ch. VI, note 6
Costelac, Tony, mine union organizer, 145
Cromwell, Oliver, 60
Crossley, Tom, forest ranger, 156
Cunningham, C. B., detective, 73
Cunningham, Dan, company guard, 21
Curtis, Vice President Charles, Kansas, 137
Cutlip Siding, West Virginia, 57

Dadisman, W. T., union official, 92
Dalhart, Texas, 160
Danville, West Virginia, 90, 92-95
Darwin, Charles, 48, 82
Daugherty, Miles, U. M. W. organizer, 43-44
Davis, Maj. Thomas B., acting adjutant general of West Virginia, 101
Davis Creek, West Virginia, 1, 142
Dawes, Vice President Charles G., 129
Dawson Coal Mining Company, 147
Delmonico's, 156
Democratic party or Democrats, 59, 128-30, 144
Denver, Colorado, 167, Ch. II, note 6
Denver and Rio Grande Western Railroad, 157
Depression, SEE Great Depression
Deputy Sheriffs, 64, 95, 97-98, 100, 104, 110, 120-21, 129
Descartes, 82
Detectives, SEE Company guards
Diana, James, U. M. W. worker, 53
Dictograph, 119, 121
Diehl, Walter, miner, 21, 44
Dillon, Harvey, miner, 92-93
Dingess Run, West Virginia, 96
Donahue, Rt. Rev. Bishop P. J., 20, 167, Ch. II, note 11
Donahue Commission, 167, Ch. II, note 11
Donwood, West Virginia, 40
Douglas, Arizona, 137, 139-40, 154
Droddy Creek (Drawdy), West Virginia, 92, 172, Ch. VIII, note 6
Dry Branch, West Virginia, 17, 167, Ch. II, note 7
Dwyer, Lawrence, U. M. W. leader, 45-46, 51-52, 169, Ch. V, note 1

Elkhorn coal field, Kentucky, 169, Ch. V, note 1
Elkins, West Virginia, x, 128
Ellis, Havelock, 82
El Paso, Texas, 132-33, 137, 140, 150, 154, 160-62
El Paso County, Texas, 160
El Paso County Jail, Texas, 160-61
El Paso County Political Alliance, 160
Engels, Friedrich, 48
Eskdale, West Virginia, 26-28, 45, 51
Estep, Cesco, miner, 36-37
Estep, Mrs. Cesco, 36-37

Fabens, Texas, 160
Fairbank, Arizona, 139
Fairmont, West Virginia, x, 58, 67, 125, 142, 144-47, 150, 171, Ch. VIII, note 1
Fairmont Hotel, 147
Fayette County, West Virginia, 123-24
Fayetteville, West Virginia, 95, 123-24, 127-28
Federal troops, 86, 91-96, 98
Felts, Albert, 73-76
Felts, Lee, 73-76
Felts, Thomas L., 87
Few Clothes, 29-31
Flemington, West Virginia, 58, 142-44
Florida, x, 131
Forestry Service, 154-58
Fort Huachuca, Arizona, 139
Fort Worth, Texas, 131
Fox, Chris P., 161
Fred's Tourist Camp, Wellton, Arizona, 133
Frey, John P., union leader, 80

Gadsden Hotel, Douglas, Arizona, 137
Gailey, S. C., coal mine manager, 41
Gallup, New Mexico, 131, 153-54
Garrison, Montana, 157
Gatens, P. F., U. M. W. leader, 45, 56
Gaujot, mine guard, 32-33
Georgia, 30
Germany, 139
Germer, Adolph, union leader, 142, 146-47, 173, Ch. XII, note 2
Gibbs, J. H. (Peggy), Hartford City, 100
Gila River, Arizona, 133
Gillespie, Illinois, 173, Ch. XII, note 4
Gilmer, West Virginia, 56-57
Gilmer County, West Virginia, 56
Gilmer Fuel Company, 57
Glasscock, Governor William E., West Virginia (1909-13), 18, 166, Ch. II, note 3
Glendale, California, 153
Goettman, C. E., lawyer, 115
Gompers, Samuel, 80, 83-84
Gore, John, deputy sheriff, Logan County, 96, 123
Gorky, Maxim, 112
Grace, Hotel, Abilene, Texas, 132
Grafton, West Virginia, x, 67, 145, 158
Grand Junction, Colorado, 158
Great Depression, 140-41, 144-45, 150, 153, 160-61
Great Lakes, 105
Greece, 60
Greeley, Horace, 130
Greenbrier County, West Virginia, 172, Ch. X, note 2
Greenway, John, union leader, 80
Griffith, C. C., U. M. W. leader, 44
Grimm, W. B., sheriff of Harrison County, 147-48
Guards, SEE company guards and gunmen
Gunmen, 14-15, 17-18, 23, 25-29, 31-35, 55-57, 59-60, 64-66, 72, 75, 88, 90-91, 94-98, 101, 104-07, 112-16, 117-18
Guyandotte River, West Virginia, 96-97

185

Haggerty, Thomas, U. M. W. leader, 41-43
Hall, John, deputy sheriff, Mingo County, 104-05
Hansford, West Virginia, 33
Haptonstall, M. L., U. M. W. leader, 109
Harding, President Warren G., 90-91, 103
Hargrove, George, U. M. W. leader, 46
Harpers Ferry, West Virginia, 123
Harqua Hala Desert, 133
Harris, J. E. "Arizona Jack", 139, 154
Harris, Rev. W. S., 9
Harris, William, labor leader, 119, 121
Harrison County, West Virginia, 147
Haskins, Jeb, 3
Hassayampa Hotel, Prescott, Arizona, 134
Hatfield, Anderson, 76-77
Hatfield, Mrs. Sid, 87-88
Hatfield, Sid, Chief of police, Matewan, West Virginia, Illustrations, 72-77, 87-89, 99, 103, 108, 170, Ch. VI, notes 11-13; 172, Ch. VIII, notes 2-4
Hawaii, 126
Hayes, Frank J., U. M. W. leader, 60
Haywood, William D., I. W. W. leader, 40, 168, Ch. IV, note 1
Haywood Junction, West Virginia, 147
*Hell before Death*, 9
Helper, Utah, 157-58
Hendricks, Mr., 137, 161-62
*Heralds of the Morning*, 7
Hereford, Arizona, 139
Hernshaw, West Virginia, 90, 92, 95
Heron, Montana, 155-56
Hickey, G. C., 123-24
Hill, New Mexico, 161
Hines, Tom, guard, 17-18
Hitchman Coal and Coke Company, ix
Hoboes, 153-54, 157-60
Hocking Coal Company, 41
Hodge, 17, 167, Ch. II, note 7
Holly Grove, West Virginia, 25-26, 28, 31-38, 72
Holstein, Edgar, 117
Holt, Homer A. "Rocky", Governor of West Virginia (1937-41), Attorney General of West Virginia (1933-37), 128
Homestead, Pennsylvania, 27
Hoover, President Herbert, 137, 140-41, 150, 161
Hoquiam, Washington, 153
Houston, Harold W., lawyer, 91, 99, 103-04, 115-16
Howatt, Alexander, labor leader, 142, 173, Ch. XII, note 4
Hughes Creek, West Virginia, 98
Hugo, Victor, 48
Hunt, Governor George W. P., Arizona, 132-33, 138
Huntington, West Virginia, 20, 35, 97, 110, 116, 118, 120
Hurst, Frank P., sheriff of Logan County, 169, Ch. VI, note 2

Idaho, 153-54, 157
Illinois, 44, 148, 173, Ch. XII, note 4
Illinois Federation of Labor, 173, Ch. XII, note 2
Immigrants, 60-62, 65
Indiana, vii, 44
Indianapolis, 102, 122, 128

Indians, 136-37
Industrial Workers of the World (I. W. W.), 168, Ch. IV, note 1
Injunctions, ix
Investigations, 20, 23, 87, 167, Ch. II, note 11; 169, Ch. VI, note 3; 179
Irishmen, 46
*Iron Heel,* 7
Italians, 49, 52, 65
Italy, 60

Jails, 101-18
Jarrolds Valley, West Virginia, 69
Jefferson, Mr., 155
Jefferson County, West Virginia, 119-21
Jeffries, Rev. J. J., 111, 113, 117
Johnson, General Hugh S., 128, 164
Johnson, Senator Hiram, California, 87
Johnson, William H., labor leader, 80
Jones, Mary Harris "Mother", 21-22, 27-28, 43, 58, 68-69, 79-85, 89-91, 132, 168, Ch. II, note 12
Jones, Mr., Coal operator, 143
Jones, Sam, revivalist, 7, 108
Joy loader, 165
Juárez, Mexico, 150, 160
*Jungle, The,* 7

Kanawha Coal Operators Association, 69
Kanawha County, West Virginia, vii-ix, 1, 59, 104, 106, 129-30, 166, Ch. I, note 3; 169, Ch. VI, note 1
Kanawha County Jail, 106-10
Kanawha River, West Virginia, 14, 31, 44
Kanawha Valley, West Virginia, 56, 89, 164
Kangaroo court, 107, 109
Kayford, West Virginia, 53, 142
Keeney, C. Frank, U. M. W. leader, Illustrations, viii-x, 27, 41, 43-46, 51-56, 63-64, 66-68, 70-71, 89-94, 98-102, 104-18, 124-28, 142-43, 146-47, 172, Ch. IX, note 1; 173, Ch. XII, note 1
Kentucky, 105, 119, 169, Ch. V, note 1
Kenyon, Senator William S., Iowa, 23
Kern, Senator John W., Indiana, 22
Kirkpatrick, James, deputy sheriff, 87-88, 101, 172, Ch. VIII, note 3
Knights of Columbus, 49
Knotts, Mr., 138
Knox, Mr., 41
Kodiak Island, 151
Kofa, Arizona, 133
Ku Klux Klan, ix, 130
Kump, Governor H. G., West Virginia (1933-37), 128

LaFollette, Senator Robert M., Wisconsin, 129
Lamar, Colorado, 158
Laredo, Texas, 79
Las Vegas, Nevada, 153
Lavinder, A. D., U. M. W. leader, 101-02, 125
Law and Order League, 130
Lens Creek, West Virginia, 90-92
Lewis, Arthur M., 8

Lewis, John L., U. M. W. president, Illustrations, x, 127-28, 142-43, 173, Ch. XII, notes 1, 4
Lewis, Kid, 158-60
Lewisburg, West Virginia, 122-23, 172, Ch. X, note 2
Lick Creek, West Virginia, 77-78, 171, Ch. VIII, note 1
Lind, West Virginia, 171, Ch. VIII, note 1
Lively, C. E., detective, 72, 88, 171, Ch. VI, note 14; 172, Ch. VIII, note 4
Logan, West Virginia, 95-97, 117, 120, 170, Ch. VI, note 6
Logan coal field, vii
Logan County, West Virginia, 63-64, 90, 93, 95-98, 101, 103-04, 110-20, 123, 164, 169, Ch. VI, note 2
Logan County Jail, 108, 110-18
London, Jack, 7, 48
Longacre, West Virginia, 41
Long Beach, California, 133
Los Angeles, California, 140, 153
Ludlow, Colorado, 72, 170, Ch. VI, note 9
Lumber industry, 5, 6
Lumberport, West Virginia, 58
Lusk, Charles, miner, 41, 45

McCartney, Frank, union organizer, 145
McCoy, Ace, 101-02
McDowell County, West Virginia, 63-64, 87, 97, 103
McGrady, Edward F., union leader, 80
McKee Construction Company, 133, 137
Madison, West Virginia, 90, 95-96
Magna Carta, 164
Manhattan, New York, 22
Marcy, 48
Marion County, West Virginia, 146
Marks, Ira, U. M. W. leader, 59, 62
Markus, Charles E., 47
Marmet, West Virginia, 65, 90-92
Martial law, 14, 16, 18-22, 86, 89, 91-95, 98, 101, 166, Ch. II, note 3; 167, Ch. II, note 9
Martine, Senator James E., New Jersey, 23
Marx, Karl, x, 48
Maryland Coal Company, 145, 147, 162
Mason, James M., Jr., Jefferson County lawyer, Illustrations, 120-21
Matewan, West Virginia, 71-77, 103
Matewan Massacre, ix, 72-77, 115, 170, Ch. VI, notes 11-14
*Menace*, 48
Messer, Mr., jailor of Mingo County, 101-02
Mexican labor, 161
Mexican Revolution, 79
Mexican visit, 79-85, 171, Ch. VII, note 6
Mexico, 79-85, 126, 132
Mexico City, 79
Middle West, 169, Ch. VI, note 1
Miley, Frank, U. M. W. leader, 143, 148
Military court, 20
Mills, Walter Thomas, 7
*Miners' Herald*, 39-40
Miners' march of 1919, ix, 65-67, 164, 170, Ch. VI, note 6
Miners' march of 1921, ix, 90-100, 164

188

Mingo County, West Virginia, 63-64, 71-73, 76-77, 86-87, 90, 97, 99-101, 104, 118, 171, Ch. VIII, note 1
Mingo County Defense League, 99, 126-27
Mingo County Jail, 101-06
Minturn, Colorado, 159
Missoula, Montana, 156-57
Missouri, x
Missouri Theatre, 136
Mitchell, Fulton, deputy sheriff, Logan County, 98
Monongah, West Virginia, 60
Montana, 155-57
Montgomery, Samuel B., Kingwood lawyer, Illustrations, 88-89, 99, 116
Montgomery, West Virginia, 39-40, 44, 67
Montgomery Ward, 138
Moody, Dwight L., revivalist, 7
Mooney, Fred, Illustrations, birth, 1; parents, 1-7, 50, 135-36; education, 1, 7-9, 47-48, 82; religion, 1-5, 7, 48-49, 85; as a coal miner, 1, 8-50; works in lumber camp, 5-6; as an orator, 8, 10; first union activity, 8-9, 12-13, 39-50; marriage, 9-10; moves to Cannelton, 14-15; in mine conflict of 1912-13, 20-22, 25-34; opposes District 17 leaders, 39-45; stabbed, 44-45; blacklisted, 46-47; kills intruder, 49, 52; wife dies, 51; elected secretary-treasurer of District 17, 51-53; as secretary-treasurer of District 17, 54-130; Mexican trip, v, 79-85; indicted, 99, 172, Ch. VIII, note 8; in Mingo County Jail, 101-06; in Kanawha County Jail, 106-10; on jails, 110; in Logan County Jail, 110-18; indicted for treason, 116; free on bail, 117-18; removed from union position, 128; candidate for House of Delegates, 129-30; on politics and politicians, 22-23, 83, 86, 107-08, 129-30, 135-36; works as carpenter, 131, 150ff; remarries, 131; divorced, 131; in Florida, x, 131; operates restaurant in Charleston, 135; and Reorganized United Mine Workers, 142-49; remarries, 150; in Far West, x, 131-35, 136-41, 150-62; abducted, 150-52; as a hobo, 153-54, 157-60; suicide, x, 186, Introduction, note 1
Mooney, Thomas J., labor leader, 81-82, 171, Ch. VII, note 3
Moore, John "Jock", 99, 103
Morgan, Governor E. F., West Virginia (1921-25), 86, 101, 103, 108, 123, 171, Ch. VIII, note 1
Morgantown, West Virginia, x, 18
Morris, B. F., U. M. W. leader and company representative, 43, 69-70
Morris Whaley, 165
Morton, Haydon, coal operator, 70
Mounds, 3
Moundsville, West Virginia, 20
Moundsville Penitentiary, 123
Mucklow, West Virginia, 25-26, 31, 33-34, 168, Ch. III, note 2

National Recovery Administration (N.R.A.), 128, 164
Naugatuck, West Virginia, 171, Ch. VIII, note 1
Needles, California, 153-54
Neely, Senator Matthew M., West Virginia, 144
Negroes, 65, 109
Nelson, Mr., 140-41
Nevada, 153
Newcomer, T., deputy sheriff, 100, 104
New Deal, 128-29, 164-65
New Mexico, x, 131, 136-37, 153, 160-61
New River, West Virginia, 164, 169, Ch. V, note 1; 172, Ch. X, note 4
New River coal field, 166, Ch. I, note 3

Newspapers, 16, 18, 23, 39-40, 63, 99, 107, 119, 122
New York, N. Y., 22, 88, 119, 122
Nolan, West Virginia, 171, Ch. VIII, note 1
Norfolk and Western Railroad, 77, 86, 105, 167, Ch. II, note 6
North Carolina, 30, 75
Northern Pacific Railroad, 153
Northern West Virginia Coal Operators Association, 172, Ch. X, note 3
Nuevo Laredo, Mexico, 79-80

Obregón, General Alvaro, 80, 85, 171, Ch. VII, note 2
O'Bryan, Captain, 103
Ohio, vii, 44, 59
*Oil*, 138
Olcott, West Virginia, 9
Oratory, 8, 10
Osenton, C. W., Fayetteville lawyer, 115-16, 120, 122

Paint Creek, West Virginia, viii, 14-39, 58, 164
Pan-American Labor Congress, January, 1921, 79-85
Payne, Forest, 117-18
Peabody, Henry, 136
Pelly, 2-4
Pennsylvania, vii, 43-44, 169, Ch. VI, note 1
Perryville, West Virginia, 42
Petry, William, U. M. W. leader, Illustrations, 53-54, 107
Peytona, West Virginia, 92
Phaupp, mine guard, 33
Phelps Dodge Corporation, 140-41
Philippine Islands, 93, 126
Phoenix, Arizona, 132-34, 137-38, 154
Pine City Tourist Camp, 134
Pinellas County, Florida, 131
Pinson, A. C., sheriff of Mingo County, 101, 104-05
Pittsburgh, Pennsylvania, 27, 94, 120, 146
Plato, 48
Pocahontas County, West Virginia, 63
Pocatello, Idaho, 157
Poland, 60
Politicians, 22-23, 63, 83, 86, 107-08, 129-30, 135-36
Portland, Oregon, 153
Powderly, T. V., 132
Pratt, West Virginia, 20, 22, 35
Prescott, Arizona, 134-35
Preston County, West Virginia, 57, 169, Ch. V, note 1
*Prince Hagen*, 7
Progressive Miners' Union, 163-64, 173, Ch. XII, note 4
Public Works Administration, 165
Pueblo, Colorado, 159
Pueblo Indian Reservation, 136-37
Purcell Walnut Company of Kansas City, 139

Railroad Brotherhoods, 119
Raleigh County, West Virginia, 53, 67
Ray, W. F., U. M. W. leader, 44-45, 56
Reagan, Mr., 139
Redbeard, 82, 171, Ch. VII, note 6
Red Necks, 31, 168, Ch. III, note 4

Red Warrior, West Virginia, 55, 169, Ch. V, note 2
Reese, W. F., U. M. W. leader, 53
Religion, 1-5, 7, 49-50
Reorganized United Mine Workers, 142-49, 173, Ch. XII, note 1
Republican party or Republicans, 18, 129-30, 171, Ch. VIII, note 1
Revivalism, 4-5
Rhonda, West Virginia, 167, Ch. II, note 7
Richmond, Virginia, 167, Ch. II, note 6
Riley, James, labor leader, 63, 99
Rio Grande River, 151, 160
Ritz, Judge Harold A., West Virginia Supreme Court, 126
Roanoke, Virginia, 15, 105
Robinson, Judge Ira E., West Virginia Supreme Court, 20
Rocky Mountains, 134
Roman Catholic Church, 48-49
Roosevelt, President Franklin D., 128
Roosevelt, President Theodore, 166, Ch. I, note 3
Roswell, New Mexico, 137
Rowan, Miss Virginia, 150
Ruffner, Hotel, Charleston, West Virginia, 43
Rumania, 60
Russia, 60
Russian Revolution, x

"St. Albans," preacher, 109
St. Albans, West Virginia, 67, 95
St. Clair, company guard, 102, 106
St. Louis, Missouri, 150
St. Petersburg, Florida, 131
St. Petersburg (Florida) Central Labor Union, x
*St. Petersburg (Florida) Union*, x
Salem, Ohio, 67
Salt Lake City, Utah, 157
Salvation Army, 153
San Carlos Hotel, 138
San Diego, California, 151
Sandpoint, Idaho, 153, 156-57
San Francisco, California, 140, 157, 171, Ch. VII, note 3
San Pedro River, Arizona, 139-40
Santa Fe, New Mexico, 136-37
Santa Fe Railroad, 153-54
Scotch Hill, West Virginia, 145
Scott, B. A., U. M. W. leader, 70
Scott, Ed, 53-56, 68-69
Scott, Isaac, 109
Scott, Mrs., 48
Scott's Run, West Virginia, 146
Scrip, 11
Seaboard Bond and Mortgage Company, Long Beach, California, 133
Seattle, Washington, 153
Sharples, 95
Shelby, Russell, miner, 57
Sheltering Arms Hospital, 33
Shields, Senator John K., Tennessee, 23
Shinnston, West Virginia, 145
Shultz Construction Company, 136
Shuttleworth brothers, company guards, 59

Sierra Vista Park, Santa Fe, New Mexico, 137
Sinclair, Upton, 7, 138
Slater, Dan, company guard, 27
Smith, 2
Smith, Howard, 21
Snyder, Frank W., 63, 99, 119, 121
Socialism or socialists, x, 20, 40, 173, Ch. XII, note 2
Sonora, Mexico, 139
South Carolina, 30
Southern Pacific Railroad, 133, 153, 157
Spanish-American War, 5, 16
Spanish-American War veterans, 92
Spencer, Herbert, 48, 82
Spokane, Washington, 153, 155
Sprigg, West Virginia, 71, 171, Ch. VIII, note 1
Springfield, Illinois, 173, Ch. XII, notes 1, 4
Squaw Mountain, Arizona, 137
Stanaford Mountain, West Virginia, 21, 72
Stollings, John, 114
Stollings, West Virginia, 96
Stomp, Charlie, miner, 30
Stone, Warren E., president of railroad engineers, 119
Stone Mountain, West Virginia, 71, 73
Strikebreakers, ix, 29-31
Strike of 1902, 166, Ch. I, note 3
Strike of 1912, 166, Ch. I, note 3
Strikes, viii, 14-39, 56-58, 71-127
Stringer, mine guard, 33
*Struggle for Existence,* 7
Sue, Eugène, 82, 171, Ch. VII, note 5
Sullivan and Jeffrey machines, 165
Sullivan Company, 131
Swanson, Senator Claude A., Virginia, 23

Talmadge, Dr., 7
Taylor County, West Virginia, 57, 162
Tenmile Fork, West Virginia, 169, Ch. V, note 2
Tent colonies, 16, 25, 29, 31, 33, 35-37, 56, 77-78, 171-72, Ch. VIII, note 1;
    SEE ALSO Holly Grove
Testerman, Cabbell, mayor of Matewan, West Virginia, 73-76, 170, Ch. VI,
    note 13
Teti, Tony, 147
Tetlow, Percy, U. M. W. leader, 67, 125, 128, 172, Ch. X, note 3
Texas, x, 131-33, 150, 154, 160-61
Thacker, West Virginia, 171, Ch. VIII, note 1
Thompson, 111-12
Thompson, Thomas, 111-12
Thompson, William, U. M. W. leader, 128
Thurmond, West Virginia, 167, Ch. II, note 6
Tittle, Charles, U. M. W. leader, 45
Toffelmire, Mr., 137
Tombstone, Arizona, 80
Topock, Arizona, 154
Townsend, T. C., Charleston lawyer, 104, 115
Treason indictments, 116
Treason trials, 98, 116, 119-26
Trinidad, Colorado, 160

Trobridge, Pike, deputy sheriff, 104
Tug River, West Virginia and Kentucky, 105

Ulm Creek, Idaho, 155-56
Union campaign in northern West Virginia, 54-63; in southern West Virginia, 63-66, 68, 71-72, 86, 89-90
Union Pacific Railroad, 153, 157
United Garment Workers, 119
*United Mine Workers' Journal*, 157, 168, 174, *passim*
United Mine Workers of America, 72, 83, 86, 130, 162-64; activity in southern West Virginia, 1902, 8-9; strike of 1912-13, 15-17, 20-22, 25-38; settlement with operators, 39; District 17, vii-ix, 39-45, 51-72, 77-79, 84, 86-87, 89-100, 103, 107, 109-10, 119, 124-26, 127-28, 158; District 17, sub-districts, 67, 69, 92, 107, 124-25; District 29, 169, Ch. V, note 1; District 30, 45-46; District 31, 148; internal conflicts or divisions, 39-46, 52-55, 58, 64, 67-69, 70-71, 90, 98, 109-10, 124-25, 128, 142-49, 163-64; elections, 51-53; campaign in northern West Virginia, 54-63; campaign in southern West Virginia, 63-66, 68, 71-72, 86, 89-90; strike of 1920-21, 71-78, 86-100, 124-26; conventions, 39-44, 102-03, 119, 122, 150
United States Army, 65, 96, 98
United States Constitution, 60, 120
United States Department of Labor, 132
United States Secretary of Labor, 44
United States Senate, Committee on Education and Labor, 22-23, 39, 87
United States Senate, Committee on Interstate Commerce, 172, Ch. X, note 3
United States Supreme Court, ix
University Heights, New Mexico, 131
Utah, 157-58

VanFleet, C. J., lawyer, Illustrations, 87, 103, 116
Villareal, Secretary, 80
Virginian Railway, 167, Ch. II, note 6
Volney, 82
Voltaire, 48

Wages, 1, 5, 24, 141
Waldo Hotel, Clarksburg, 148
Walker, Capt. S. L., 167, Ch. II, note 11
Walker, H. A., sheriff of Kanawha County, 104
Walker, John H., union leader, 142, 173, Ch. XII, note 2
Wallace, George S., judge advocate of West Virginia, 20
Ware Ramey Company, 137
Washington, 153-55
Washington, D. C., 87, 122
Washington Hotel, Charleston, 122
Watkins, Andrew, 59
Watson, Clarence, 60
Weir, Mr. and Mrs. James W., 65
Weirwood, West Virginia, 97
Weiser, Idaho, 154
Welch, West Virginia, 76, 87, 103
Wellton, Arizona, 133
Wendel, West Virginia, Illustrations, 145, 162-63
*West Virginia Federationist*, 63, 99, 119

West Virginia Federation of Labor, SEE West Virginia State Federation of Labor
West Virginia House of Delegates, 129-30
West Virginia State Capitol, 22, 89
West Virginia State Federation of Labor, 63, 99, 119
West Virginia State Militia, 18-19
West Virginia State Police, 95, 170, Ch. VI, note 12
West Virginia Supreme Court, 20, 86
Wheeling, West Virginia, 20, 146
"When the Leaves Come Out", poem, 38
White, "Dad", jailor and deputy sheriff of Logan County, 94, 110, 113
White, John P., U. M. W. leader, 168, Ch. IV, note 2
White, Lewis, 94, 96
White, Mr., 133
Whitesville, West Virginia, 68-69
Wilburn, Isaac, 96
Wilburn, John, 96, 123
Wilburn, Rev. J. E., 96, 123
Williams, R. M., U. M. W. field worker, 99
Williams, Arizona, 154
Williamson, David, coal operator, 162
Williamson, West Virginia, 77-78, 100-01, 103-05
Wilson, President Woodrow, 80
Wilson, William B., U. S. Secretary of Labor, 44
Winebrennarian Church, 3
Winifrede, West Virginia, 93
WMMN, Radio Station, Fairmont, West Virginia, 146
Wolcott, Colorado, 158-59
Woods, William, city marshal, Carlsbad, New Mexico, 137
Workman, C. H., U. M. W. leader, 53, 57, 101-02
World War I, viii-ix, 169, Ch. VI, note 1
World War I veterans, 65, 92
Wright, Chester M., union leader, 80
Wyoming County, West Virginia, 63

Youngstown, Ohio, 27
Yuma, Arizona, 133
Yuma County, Arizona, 133

Zuni Mountains, New Mexico, 131